AF531678

ENCYCLOPAEDIA OF MOLECULAR BIOLOGY

ENCYCLOPAEDIA OF MOLECULAR BIOLOGY

(Vol. 2)

A. Tyagi

ANMOL PUBLICATIONS PVT. LTD.
NEW DELHI - 110 002 (INDIA)

ANMOL PUBLICATIONS PVT. LTD.
H.O.: 4374/4B, Ansari Road, Darya Ganj,
New Delhi-110 002 (India)
Ph.: 23261597, 23278000
B.O.: No. 1015, Ist Main Road, BSK IIIrd Stage
IIIrd Phase, IIIrd Block,
Bangalore - 560 085 (India)
Visit us at: www.anmolpublications.com

Encyclopaedia of Molecular Biology

First Published, 2006

ISBN 81-261-2677-9 (Set)

[Responsibility for the facts stated, opinions expressed, conclusions reached and plagiarism, if any, in these volumes is entirely that of the Author. The Publishers bear no responsibility for them, whatsoever.]

PRINTED IN INDIA

Published by J.L. Kumar for Anmol Publications Pvt. Ltd., New Delhi - 110 002 and Printed at Mehra Offset Press, Delhi.

Contents

Preface

Smallest fundamental unit of an element or a compound is called a molecule. The rapid progress and multi-faceted development of Science, in general; and of Biology, in particular, has ushered in an absolutely new era in the history of mankind, regarding the physio-chemical department of life.

There has been an urge to know more and more from the day one. It is due to this urge that man has been scaling ever new heights day-after-day, be it related to Science or Commerce or Politics and so on. Is this whole endeavour, which has proved to be a real boon for man, or in other words, the greatest gift, the twentieth century gave to humanity ? It really brought about a revolution in the whole human life. The emergence of Molecular Biology, as a branch of Science for study, could also be possible by virtue of this very present of the past.

The discipline is very vast and with the passage of time, big breakthroughs are in the offing. Feeling the pulse of the time, it deemed suitable, rather, necessary to bring out a comprehensive and compact encyclopaedia on this subject in order to cater to the maximum requirements of the lovers of Molecular Biology.

Comprising five volumes, this trivial attempt is expected to prove to be an asset for Science elite and students alike. Now, savour it molecule-by-molecule.

Editor

Preface

Smallest fundamental unit of an element or a compound is called a molecule. The rapid progress and multifaceted development of Science in general and of Biology in particular, has ushered in an absolutely new era in the history of mankind, regarding the physio-chemical department of life.

There has been an urge to know more and more from the day one. It is due to this urge that man has been scaling ever new heights day after day, be it related to Science or Commerce or Politics and so on. In this whole endeavour, which has proved to be a real boon for man, or in other words, the greatest gift the twentieth century gave to humanity, it really brought about a revolution in the whole human life. The emergence of Molecular Biology as a Branch of Science for study could also be possible by virtue of this very present of the past.

The discipline is very vast and with the passage of time, big breakthroughs are in the offing. Feeling the pulse of the time, it deemed suitable, rather, necessary to bring out a comprehensive and compact encyclopaedia on this subject in order to cater to the maximum requirements of the lovers of Molecular Biology.

Comprising five volumes, this trivial attempt is expected to prove to be an asset for Science olds and students alike. Now, savour it molecule-by-molecule.

Editor

1

Genetic Code

We know that the genetic information in the nucleic acids is expressed through synthesis of proteins. We also know that, while proteins have 20 different kinds of essential amino acids, nucleic acids (particularly in living cells.) have only four different kinds of bases. (Recently, it has been shown that in artificially synthesized DNA segments, the number of bases can extend to more than four i.e. 6 or 8). It may therefore, be realized that while 20 amino acids are the alphabets of the language of proteins, four bases are the alphabets of the language of nucleic acids.

Since it is through mRNA (messenger RNA) that the genetic information is passed on to proteins, the problem of genetic code was to prepare a dictionary for translating the language of RNA into the language of proteins. Since one is four alphabet language and the other has 20 alphabets, as shown in Figure, a singlet code would give only four codons, a doublet code would have 16 codons and triplet code would give us 64 ($4 \times 4 \times 4$) triplets. The 64 triplets would be enough to code for 20 amino acids.

In a 64-codon dictionary, either there should be more than one triplets coding for the same amino acid or else, of the 64 triplets, 44 triplets should mean nothing (nonsense). It has now been proved by experimental evidence that the code is triplet and that all 64 codons carry some meaning, although three of these 64 triplets cannot be translated in terms of amino acids, since they mean stop signals. Whenever any one of these three triplets is encountered on mRNA, it orders that the end of the polypeptide chain has reached and that the chain should now be released, before further synthesis may take place.

singlet code
A
G
C
T

doublet code			
AA	AG	AC	AT
GA	GG	GC	GT
CA	CG	CC	CT
TA	TG	TC	TT

triplet code			
AAA	GAA	CAA	TAA
AAG	GAG	CAG	TAG
AAC	GAC	CAC	TAC
AAT	GAT	CAT	TAT
AGA	GGA	CGA	TGA
AGG	GGG	CGG	TGG
AGC	GGC	CGC	TGC
AGT	GGT	CGT	TGT
ACA	GCA	CCA	TCA
ACG	GCG	CCG	TCG
ACC	GCC	CCC	TCC
ACT	GCT	CCT	TCT
ATA	GTA	CTA	TTA
ATG	GTG	CTG	TTG
ATC	GTC	CTC	TTC
ATT	GTT	CTT	TTT

A singlet code, a doublet code and a triplet code.

Properties of Genetic Code

Following properties of the genetic code were proved by definite experimental evidence : (i) the code is triplet, (ii) the code is degenerate, (iii) the code is non-overlapping, (iv) the code is commaless, (v) the code is non-ambiguous and (vi) the code is universal. Although it may not be necessary to present the experimental evidences which proved the validity of these properties, it may be useful to explain the meaning of these six properties of the genetic code listed above.

The Code is Triplet : As earlier outlined, singlet and doublet codes are not enough to code for 20 amino acids; it was pointed out that triplet code is the minimum required. But it could be a quadruplet code or of a higher order. As pointed out above, in a triplet code of 64 codons, there is an excess of 44 codons and, therefore, more than one codons are present for the same amino acid. This excess will be still greater if more than three-letter words are used. In the quadruplet code there will be $4 \times 4 \times 4 \times 4 = 256$ possible words.

The Code is Degenerate : In a triplet code, as pointed out earlier also, for a particular amino acid more than one words (synonyms) can be used. This phenomenon is described by saying that the code is degenerate. A non-degenerate code would be one where there is one to one relationship between amino acids and the codons, so that 44 codons out of 64, will be useless or nonsense codons. It has been definitely shown that there are no nonsense codons. The codons which were earlier called nonsense codons are also now known to mean stop signals.

The Code is Non-overlapping : Non-overlapping code means that a base in a mRNA is not used for two different codons. In Figure it is shown that an overlapping code can mean coding for four amino acids from six bases. In actual practice six bases code for not more than two amino acids. However, in overlapping genes described earlier it is shown that the same base can be used for different codons, but only at different occasions in time and/or space, so that the same base can not be used for two different codons during synthesis

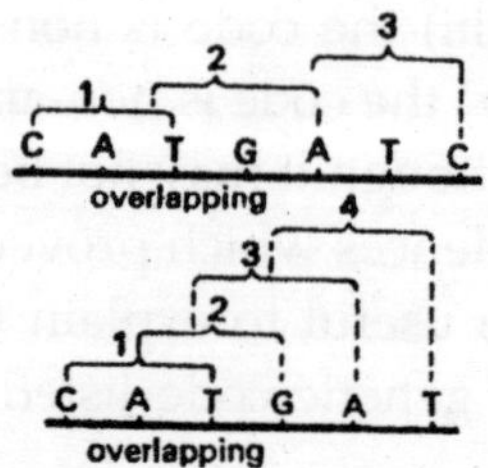

Overlapping of codones due to one letter or two letters.

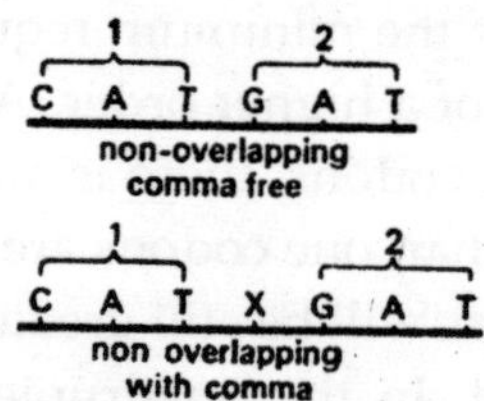

Genetic code, without comma and with comma.

of the same protein. This is exactly, what we mean by non-overlapping code.

The Code is Commaless : A commaless code means that no punctuations are needed between any two words. In other words, we can say that after one amino acid is coded, the second amino acid will be automatically coded by the next three letters and that no letters are wasted for telling that one amino acid has been coded and that now second should be coded (Fig.).

The Code is Non-ambiguous : Non-ambiguous code means that there is no ambiguity about a particular codon. A particular codon will always code for the same amino acid, wherever it is found. In an ambiguous code, the same codon could have different meanings, or in other words, the same codon could code two or more than two different amino acids. Such is not the case. While the same amino acid can be coded by more than one codons (the code is degenerate), the same codon shall never code for two different amino acids (non ambiguous).

However, there is some element of ambiguity when AUG and GUG are considered; both may code for *methionine* as initiating codons, although GUG is meant for *valine* (Fig.). Moreover, a different code exists in mitochondria of some eukaryotes, so that in cytoplasm and mitochondria same codon may code for different amino acids. This may not mean ambiguity, since in mitochondria there is a separate genetic code which differs from universal code in some essential respects.

The Code is Universal : Although the code has been worked out by using *in vitro* systems prepared usually from micro-organisms, there is no doubt now that in all kinds of living organisms, micro- or macro-, plants or animals, the same genetic code is used. However, as will be seen later in this chapter, a different and more primitive genetic code exists in mitochondria of some organisms.

Codon Assignments

Although theoretical considerations in 1950's had suggested that the genetic code should be triplet in nature, it was not possible to say which codon of the possible 64 codons should code for which of the 20 essential amino acids. The first clue to this problem came when M.W. Nirenberg (Nobel Prize winner with H.G. Khorana and R.W. Holley in 1968) used *in vitro* system for the synthesis of a polypeptide using an artificially synthesized mRNA molecule. The use of artificial mRNA molecule has a definite advantage, since one could know something about its structure.

Assignment of Codons with Unknown Sequence

Polyuridylic Acid or Poly-U and other Homopolymers : In the first place, M.W. Nirenberg and J.H. Matthaei synthesized RNA using only uracil, so that all along the length of mRNA, there was no other base and the only possible triplet was UUU. When such a poly-U (RNA) was used in cell free synthesis of a polypeptide (using cell extracts from *E. coli* and supplying all the required components of protein synthesizing machinery), only polyphenylalanine was synthesized, meaning that the only amino acid coded was phenylalanine. It was, therefore, immediately concluded that the triplet UUU coded for the amino acid *phenylalanine.* This news that the genetic code had been cracked splashed in the newspapers all over the world in 1961. Subsequently, poly A gave *polylysine* and poly C gave *polyproline.* Therefore, AAA was assigned to *lysine* and CCC to *proline.* Similar experiments with poly G were not successful, because it attains secondary structure and thus cannot attach to ribosomes. The following three codons thus could be assigned, using homopolymers of RNA, without any difficulty; UUU = *phenylalanine;* AAA = *lysine;* CCC = *proline.*

Copolymers : Having successfully used the homopolymers, Nirenberg and his co-workers used RNA synthesized by using two or more bases. For instance, if only A and C are used, poly AC will consist of eight possible codons namely AAA, AAC, ACA, CAA, CCA, CAC, ACC and CCC. The proportion of these eight codons in the synthetic RNA can be calculated if the known quantities of A and C are used for the synthesis of poly AC. For instance, if A : C = 5 : 1, (5/6 is A and 1/6 is C), the calculated relative proportions of eight codons on random basis would be as given in Table.

The Relative Proportions of Different Codons in mKNA Formed Due to Bases A : C taken in 5 : 1 Ratio.

Base composition	*Codon & probability*	*Ratio using*	
		min. (as 1)	*max. (as 100)*
(1) 3A.	1. AAA 5/6 × 5/6 × 5/6 = 125/216	125	100
(2) 2A1C	2. AAC 5/6 × 5/6 × 1/6 = 25/216	25	20
	3. ACA 5/6 × 1/6 × 5/6 = 25/216	25	20
	4. CAA 1/6 × 5/6 × 5/6 = 25/216	25	20
(3) 1A2C	5. CCA 1/6 × 1/6 × 5/6 = 5/216	5	4

Contd...

Base composition	*Codon & probability*	*Ratio using* min. (as 1)	*Ratio using* max. (as 100)
	6. CAC 1/6 × 5/6 × 1/6 = 5/216	5	4
	7. ACC 5/6 × 1/6 × 1/6 = 5/216	5	4
(4) 3C	8. CCC 1/6 × 1/6 × 1/6 = 1/216	1	0.80

The calculated relative proportions of codons were compared with the proportions in which different amino acids were present in the polypeptides synthesized using poly AC. For instance, if an amino acid is 1/5th of lysine (coded by AAA), we can say that it should be coded by one of the three possible 2A1C codons (AAC, ACA, or CAA).

Similar reasoning would allow assignments of 1A2C as well as 3C. However, using this technique, it was not possible to assign the three codons of the category 2A1C (i.e. AAC, ACA. CAA) to three amino acids, since these will be present in equal quantities. Therefore, the codons were initially assigned only with respect to base composition, ignoring the sequences of the bases in codons, as done in the above example. The assignments are given in Table.

Codon Assignments Derived Due to Use of A : C (5 : 1) in the Synthesis of mRNA.

Amino acids	*Codon composition*
Lysine	3A
Asparagine, glutamine & threonine	2A1C
Histidine, proline & threonine	1A2C
Proline	3C

Assignment of Codons with Known Base Sequence : A large number of synthetic copolymers of unknown sequences was used for preparing a dictionary where codon compositions were assigned to amino acids. The next problem and a serious one was to sort out the three codons in one base composition e.g. 2A1C or 1A2C. In other words, we can say that the next problem was to work out the sequences of bases in codons, whose composition was already known. This was done mainly by the following methods.

Binding technique of Nirenberg and Leder. M.W. Nirenberg and P. Leder in 1964 found that if a synthetic trinucleotide for a known sequence (with known bases at 5' end and 3' end) is used with ribosome and a particular aminoacyl-tRNA (tRNA having its own specific amino acid attached), these will form a complex, provided the used codon codes for the amino acid attached to the given aminoacyl tRNA.

Codon 1 + Ribosome + AA_1 – $tRNA_1$ → Ribosome-$Codon_1$-AA_1-$tRNA_1$

In a process such as above, if given AA_1 is used with a given codon 1 and the formation of the complex is detected, this would prove that the given codon codes for the given amino acid.

It was also observed that while the free AA-tRNA passes through nitrocellulose membrane, the ribosome-codon-AA-tRNA complex adsorbs on such a membrane. If in a particular mixture only one of the amino acids is made radioactive, then the presence or absence of the radioactivity on the nitrocellulose membrane will show whether there is a relationship between the codon and the amino acid which was made radioactive.

For instance, 20 samples of a mixture of all 20 amino acids may be taken and in each sample one amino acid is made radioactive in such a manner that each and every amino acid is made radioactive in one sample or the other, and no two samples have same radioactive amino acid.

A particular sample would be then known by its radioactive amino acid. Now tRNAs and ribosomes are mixed with each sample and same codon is used for complex formation in all 20 cases. When the mixture is poured on the nitrocellulose membrane, radioactivity on membrane will be observed only when the radioactive amino acid is taking part in the formation of the complex.

Since in each sample the radioactive amino acid is known it would be possible to detect the amino acid coded by a given codon by the presence of radioactivity on the membrane. Such a treatment was given by Nirenberg and his co-workers to all the 64 synthetic codons, and their respective amino acids were identified. The binding of AA-tRNA was not equally efficient in all cases. Therefore, the sequences of bases in only about 45 codons could be worked out by this method.

Copolymers of repetitive sequences by H.G. Khorana. As outlined above, sufficient information regarding the sequences of bases in codons was available through the painstaking work of Nirenberg and co-workers. However, H.G. Khorana, almost at the same time, devised an ingenious technique for the same purpose.

Using synthetic DNA, Khorana and his co-workers could prepare polyribonucleotides (RNA) with known repeating sequences. A repeating sequence means that, if CU are two bases, these will be repeatedly present throughout the length as follows :

CUCUCUCUCUCUCU

In a similar manner, if ACU are three bases they will be present repeatedly as follows :

ACUACUACUACU

Such copolymers will direct the incorporation of amino acids in a manner which can be theoretically predicted. For instance, in $(CU)_n$ = (CUC/UCU/CUC/UCU), only two codons are possible and these are CUC and UCU. Moreover, these

Assignment of Codons, Having Known Sequences, with the Help of Copolymers having Repetitive Sequences of Two Bases

Copolymers used	Codons in copolymer	Amino acids incorporated	Codons assigned
$(CU)_n$	CUC/UCU/CUC	leucine/serine	CUC/UCU
$(UG)_n$	UGU/GUG/UGU	cysteine/valine	UGU/GUG
$(AC)_n$	ACA/CAC/ACA	threonine/histidine	ACA/CAC

Assignment of Codons, Having Known Sequences, with the Help of Copolymers having Repetitive Sequences of Three Bases = $(ACG)_n$.

Codons	Homopolypeptide	Codon assignment
ACG/ACG/ACG/ACG/ACG=Poly (ACG)	$(\text{Threonine})_n$	ACG = threonine
A/CGA/CGA/CGA/CGA=Poly (CGA)	$(\text{Arginine})_n$,	CGA = arginine
AC/GAC/GAC/GAC/GAC=Poly (GAC)	$(\text{Aspartic acid})_n$	GAC = asartic acid

codons are present in alternating sequence. The result would be that the polypeptide formed would have only two amino acids in alternating sequences. These two amino acids can be assigned to the two codons (Table).

We may similarly consider a repeating sequence of three bases e.g. $(ACG)_n$. Depending upon where the reading is started, three kinds of homopolypeptides are expected (Table). Actual codon assignment i.e. to find out which of the three codons codes for which amino acid would depend upon the previous information available regarding the composition of bases in different codons coding for different amino acids. This information was discussed earlier in the chapter.

On the basis of the above techniques, a complete genetic code dictionary could be prepared which is given in Figure. It can be easily noticed that there are two codons namely AUG and GUG, which are designated as initiation codons and there are three codons UAA, UAG and UGA which are designated as termination codons.

Chain Initiation and Chain Termination Codons : In most if not all polypeptides synthesized *in vivo,* the initiating amino acid is methionine. The corresponding codon on mRNA is usually AUG, but rarely it could be GUG also. When GUG is used as the initiating codon, it does not code for

First letter	Second letter: U		C		A		G		Third letter
U	UUU	Phe (F)	UCU	Ser (S)	UAU	Tyr (Y)	UGU	Cys (C)	U
	UUC		UCC		UAC		UGC		C
	UUA	Leu (L)	UCA		UAA	stop	UGA	stop SeCys	A
	UUG		UCG		UAG	stop	UGG	Trp (W)	G
C	CUU	Leu (L)	CCU	Pro (P)	CAU	His (H)	CGU	Arg (R)	U
	CUC		CCC		CAC		CGC		C
	CUA		CCA		CAA	Gln (Q)	CGA		A
	CUG		CCG		CAG		CGG		G
A	AUU	Ile (I)	ACU	Thr (T)	AAU	Asn (N)	AGU	Ser (S)	U
	AUC		ACC		AAC		AGC		C
	AUA		ACA		AAA	Lys (K)	AGA	Arg (R)	A
	AUG*	Met (M)	ACG		AAG		AGG		G
G	GUU	Val (V)	GCU	Ala (A)	GAU	Asp (D)	GGU	Gly (G)	U
	GUC		GCC		GAC		GGC		C
	GUA		GCA		GAA	Glu (E)	GGA		A
	GUG*		GCG		GAG		GGG		G

Genetic code dictionary; three letter and one letter abbreviations are used for amino acids (one letter abbreviations are given in parentheses).

valine (as shown in Fig.). However, initiating methionine molecule should be found in the formylated state. This makes a distinction between the initiating methionine and the methionine at internal position. The methionine when required at internal position should not be formylated. Also while formyl methionine is carried by $tRNA_f^{met}$, there is a separate species of tRNA for internal methionine and it is designated as $tRNA_m^{met}$.

It should also be realized that although the genetic code *in vivo* is unambiguous (no codon coding for more than one amino acid) there is an element of ambiguity with GUG, since if present at initiating position, it would code for methionine, but would otherwise always code for valine.

As shown in Figure, UAA (also known as ochre), UAG (also known as amber) and UGA (also known as opal) are three codons which whenever present in mRNA, would bring about termination of polypeptide chain which would then be released from the ribosome. Till the function of these codons (UAA, UAG, UGA) was discovered, they were called nonsense codons, since these codons did not code for any of the 20 essential amino acids.

Now that we know, that they have a definite function of termination to perform, it would not be correct to call them nonsense codons.

Synonym Codons and Degeneracy

As earlier pointed out and as can be seen from Figure, there are more than one codons for each of several amino acids (see Table). The frequency of different amino acids in proteins also agrees with the number of codons representing each amino acid (Fig.), thus suggesting that the synonym codons have been retained in the genetic code for a definite purpose and is not a wasteful phenomenon.

It may also be seen that codons representing same amino acid are clustered in the genetic code dictionary (Fig.), and that the synonym codons usually differ in the third base only, a phenomenon sometimes called third base degeneracy (Table).

Mutations and the Genetic Code

Most of the work earlier discussed in this chapter deals with the study of genetic code in cell free systems. Therefore, it could be questioned whether or not this information would apply to the living systems. With the study of certain mutants,

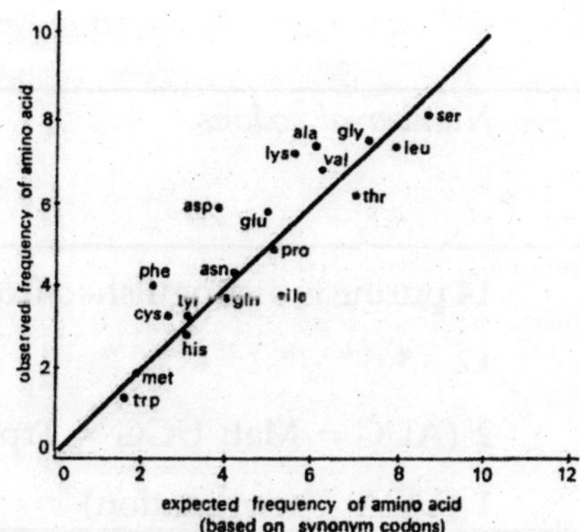

Relationships of frequencies of codons for different amino acids in the genetic code (this gives expected frequencies of amino acids in proteins) and the observed frequencies of amino acids in proteins of living systems (drawn from Lewin's "Genes VI")

Amino Acids Coded by One or More Codons in Genetic Code Dictionary.

Number of codons	*Amino acids*
One	*Met*, (initiating amino acid), *Trp*.
Two	*Phe, Tyr, His, GluN, AspN, Lys, Asp, Glu, Cys.*
Three	*Ileu.*
Six	*Ser, Arg, Leu.*

it was possible to show that the genetic code deciphered using cell free systems applies to the living systems also. There are

Frequencies of Codons having Third Base Degeneracy

Third base having same meaning	*Number of codons*
U, C, A, G	32
U, C, A	3 (Ileu)

Contd...

Third base having same meaning	*Number of codons*
U, C	14 purine is distinguished from pyrimidine
A, G	12 " " " " " " " "
G only	2 (AUG = Met; UGG = Trp)
A only	1 (UGA = termination)

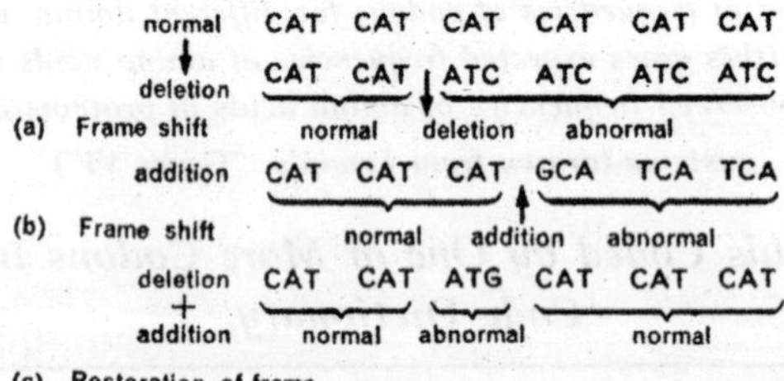

Frame-shift mutations due to deletion and addition.

two kinds of mutations which played a very significant role in the study of the genetic code in living systems. These are (i) frameshift mutations and (ii) base substitutions.

Frameshift Mutations : The genetic message, once initiated at a fixed point, is read in a definite frame in a series of three letter words. The framework would be disturbed as soon as there is a deletion or addition of one or more bases. There is evidence that mutations induced by acridine dye and related compounds depend on addition or deletion of single bases. When such frameshift mutations were intercrossed, in certain combinations they gave wild type.

It was concluded that one of them was deletion and the other an addition, so that the disturbed order of the frame due to one mutation will be restored by the other (Fig.). In such a restoration, a few amino acids coded by mRNA segment lying between the sites of deletion and addition would differ from normal protein. These differences could be explained on the basis of the Genetic Code and, therefore, are used as an evidence in favour of the properties described for Genetic Code.

Base Substitutions or Amino Acid Replacements : If in a mRNA at a particular point, one base pair is replaced by another without any deletion or addition, the meaning of one codon containing this altered base will change. Consequently, in place of a particular amino acid at a particular position in a polypeptide, another amino acid will be incorporated. Such mutations have been studied in detail in enzyme tryptophan synthetase. By working out the altered amino acid sequences in the polypeptides, some conclusions regarding the possible changes in the base-sequence in RNA can be easily made, if the dictionary of genetic code is available. This has been done in *tryptophan synthetase* and also in case of haemoglobin using sickle cell anemia.

More details about these mutations are given later in this chapter, while discussing suppressor mutations. All such mutations could be easily explained on the basis of the Genetic Code dictionary given in Figure. The mutations thus provide an evidence to suggest that the Genetic Code deciphered in cell free system holds good in living cells also.

Wobble Hypothesis

It is sometimes argued that the third base of a codon is not very important and that specificity of a codon is particularly determined by the first two bases. It has been shown, for instance that the same tRNA can recognize more than one codons differing only at the third position. This pairing is not very stable and is allowed due to wobbling in base pairing at this third position. Dr. F.H.C. Crick, in 1965 proposed a hypothesis called *Wobble hypothesis* to explain this phenomenon.

He discovered that if U is present at first position of anticodon, it can pair with either A or G at the third position of codon. Similar is the case with G and I (I = inosine is a modified base in tRNA) found in anticodon. The pairing relationships between first base of anticodon and third base of codon are given in Table).

Wobble Pairing

Anticodon (first base)	*Codon (third base)*
U	A, G
C	G
A	U
G	U, C
I (inosine, resembles G)	U, C, A

This kind of wobbling allows economy of the number of tRNA molecules, since several codons meant for same amino acid are recognized by same tRNA. For instance, anticodon IGC can recognize codons GCU, GCC and GCA (coding for alanine); IGA can recognize UCU, UCC and UCA (coding for serine) and IAC can recognize GUU, GUC and

Differences Between Universal Genetic Code and Mitochondria) Genetic Code.

Universal genetic code	*Mitochondrial code (mammals, yeast)*
1. 55 anticodons (RNA)	22 anticodons (RNA)
2. 3 termination codons (UAA, UAG, UGA)	4 termination codons (UAA, UAG, AGA, AGG)
(a) UGA = termination	UGA codes for tryptophan
(b) AGA and AGG code for arginine	AGA and AGG are termination codons
3. CUN (N = any nucleotide) codes for leucine	CUN codes for threonine in yeast mitochondria
4. UAG anticodon is aminoacylated by leucine	UAG anticodon in yeast mitochondria accepts threonine

New Genetic Codes in Mitochondria and Ciliate Protozoa : Earlier we described that the genetic code is universal. It was also believed that the genetic code does not undergo any kind of evolution and, therefore, should be static. However, during 1980s, it was shown that variations in genetic code are found in mitochondria, particularly studied in yeast and mammals. In these cases (yeast mitochondria), UGA codes for tryptophan, although in the nuclear genes, UGA is a termination codon (Fig.). It has also been shown that although a new genetic code may exist in mitochondria, but mitochondria in all organisms may not have the same genetic code. For instance, UGA in mitochondria, does not always code for tryptophan.

In the gene coding for maize cytochrome oxidase subunit II, UGA codon is not present and tryptophan is coded by codon CGG (coding for arginine in the nuclear genes; see Fig.), in addition to the codon UGG. There are also indications that AUA (coding for isoleucine in nuclear genes (see Fig.) codes for methionine in mitochondria of mammals, *Xenopus*, yeast, and *Drosophila*, but not in those of *Neurospora* and *Aspergillus*. The CUN triplets (CUA, CUG, CUU, CUC) in yeast mitochondria, but not in some fungi, code for threonine instead of leucine. In mitochondria of mammals, but not in fungi, there are four initiating codons (AUG, AUA, AUU, AUC) instead of two (AUG, GUG) in the universal code, and they are recognized by N-formylmethionyl-tRNA as in bacteria, but unlike

second base		second base		third base
C	UCU	U	UUU	U
	UCC		UUC	C
	UCA		UUA	A
	UCG		UUG	G
A	UAU	G	UGU	U
	UAC		UGC	C
	UAA		UGA	A
	UAG		UGG	G

An archetypal code, showing 16 (sixteen) anticodons, with first base (5') always U (uracil) which can wobble and pair with any of the four bases at 3' end of mRNA triplet codon—(for details see text).

eukaryotes. Another unique feature of mitochondrial code of mammals and *Xenopus,* is the use of AGA and/or AGG as stop codons rather than as arginine codons. In mitochondria of *Drosophila* AGA appears to code for serine. It is thus obvious that there is considerable diversity in the genetic code of mitochondria (Table).

The need for a different genetic code in mitochondria perhaps arose due to small genome size in mitochondria. This necessitated an evolutionary retrogression towards a more primitive code. In 1966, T.H. Jukes presented the concept of primitive code called archetypal code, in which one anticodon (in tRNA) will pair with a family of four codons due to wobbling at the first base of anticodon and third base of codon. This four way pairing will provide for only 15 amino acids instead of 20, since there are 16 anticodons (Fig.) and one of them will have to be excluded to provide for chain termination. However, in mammalian mitochondria (Mmt) this archetypal code is modified since provision needs to be made for 20 amino acids.

Sequences of complete mitochondrial DNA are now known in human, bovine, mouse and some other organisms. Each

anti codon	amino acid	anti codon	amino acid	anti codon	amino acid	anti codon	amino acid
GAA	phe	UGA	ser	GUA	tyr	GCA	cys
UAA	leu			—}stop		UCA	tryp
UAG	leu	UGG	pro	GUG	his	UCG	arg
				UUG	gluN		
GAU	ileu	UGU	thr	GUU	aspN	GCU	ser
CAU	met			UUU	lys	—}stop	
UAC	val	UGC	ala	GUC	asp	UCC	gly
				UUC	glu		

Anticodons in the mammalian mitochondrial genetic code; anticodons in each box are meant for four mRNA triplet codons shown in Fig. For instance, anticodon GAA pairs with codons UUU and UUC, anticodon UAA with codons UUA and UUG, anticodon UAG with CUU, CUC, CUA and CUG and so on.

mitochondrial DNA has been shown to have genes for 22 tRNAs as against 55 tRNAs in universal genetic code discovered in *E. coli* and applicable in most cases. These 22 tRNAs are sufficient to read codons for 20 amino acids and can pair with 60 codons, remaining four codons being used for stop signals in Mmt. This reduction in the number of tRNAs is achieved by four way pairing and two way pairing.

Four way pairing is used eight times (32 codons, 8 amino acids, 8 tRNAs); G pairs with U and C six times and U pairs with A and G another six times (24 codons; 12 amino acids; 12 tRNAs). This provides for 20 amino acids with 56 codons (using 20 tRNAs). Of the remaining eight codons, there are two each for leucine and serine using two additional tRNAs, since both of them have six codons each and there are four stop codons (UAA, UAG, AGA, AGG).

These 22 anticodons and their corresponding 20 amino acids are given in Figure. The differences between universal code and mitochondrial code are given in Table, and the diversity in mitochondrial code in several organisms is shown in Table.

It will be interesting to compare Figure with Figure. The major differences include the following : (i) one termination codon (UGA) and two arginine codons (AGA, AGG) from universal code have different meanings in the

Differences in Meaning of Several Codons in Universal Code and Mitochondrial Genetic Code in a Few Organisms.

Organism	*Codon*	*Probable meaning in mitochondrion*	*Usual meaning*
All cases	UGA	Tryptophan	Termination
Yeast	CUA	Threonine	Leucine
Fruitfly	AGA	Serine	Arginine

Contd....

Organism	*Codon*	*Probable meaning in mitochondrion*	*Usual meaning*
Mammal	AGA }	Termination	Arginine
Mammal	AGG }		
Mammal	AUA	Methionine	Isoleucine

mitochondrial genetic code; UGA in Mmt (mammalian mitochondria) coding for tryptophan and AGA and AGG in Mmt coding for stop signals; (ii) number of tRNAs is 22 in Mmt code, while it is 55 in *E. coli.* In Table it is also apparent that the mitochondrial code in mammals and yeast may differ in some respects.

In 1986, a different genetic code (like in mitochondria) has been shown to be present in ciliate protozoa *(Mycoplasma capricolum).* In this genetic code, codons UAA and UAG specify glutamine instead of stop signals. In future, more such cases may come to light among primitive organisms.

Suppression Mutations, Base Substitutions and Suppressor tRNAs : When the effect of one mutation is suppressed by another mutation, the latter is called a suppressor mutation. For instance, *AABB* may be wild, and *aaBB* a mutant; then if *aabb,* is also a wild type, the mutation in *'B'* will be called suppression mutation, because the effect of *'a'* is suppressed by *'b'*. Mechanism of this suppression effect was not known for some time, but later its mechanism became largely known.

These effects can be conveniently classified into (i) intragenic suppression, and (ii) intergenic suppression.

Intragenic Suppression : Suppression of a mutation at one site within a gene can be achieved by mutation at another site within the same gene. For instance, a mutant A46 that produces inactive tryptophan synthetase enzyme, due to substitution of glutamic acid for glycine, at position 210 may be corrected or

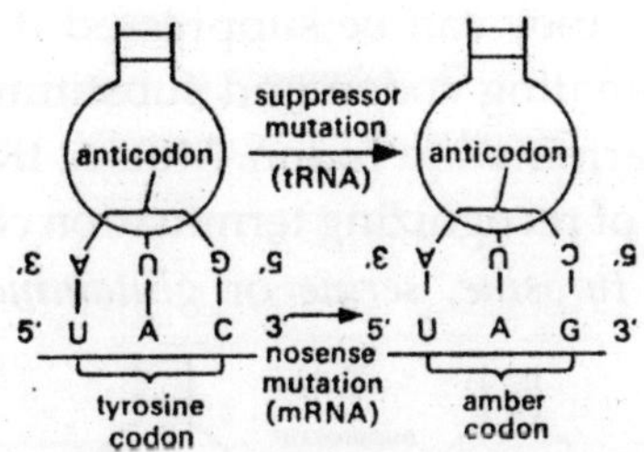

Suppressor mutation involving a change in one base of anticodon of tyrosine tRNA leading to reading of termination amber codon UAG, which resulted from nonsense mutation in tyrosine codon UAC thus leading to suppression.

suppressed due to mutation A446 involving substitution of cysteine in place of tyrosine at position 174. These two mutants A46 and A446 independently give rise to inactive enzyme, but if present together give rise to active enzyme, thus suppressing the mutant effect of each other. In many other combinations, double mutants do not show any suppression effect. The suppression effect really depends on whether or not the second alteration can restore the 3-D configuration of the enzyme. There are a number of known examples of such intragenic or intracistronic suppression or complementation. The reader is advised to consult an advanced textbook for further details in this connection.

Intergenic Suppression : Intergenic suppression, which means suppression of mutation in one gene due to mutation in another gene, can be achieved mainly due to alteration in transfer RNA (tRNA). The suppression may also be effective by alteration brought about in ribosome conformation. There are at least four types of such suppression mutations, which can be discussed.

Nonsense Suppression and Mutant tRNA : Sometimes due to a change in anticodon of a specific tRNA, it may develop the capacity to read the mutant terminating codon (amber or UAG, ochre or UAA and opal or UGA mutants). Since these nonsense mutations cause premature termination of polypeptide chain, they can be suppressed if a mutant tRNA can read the terminating codon and substitute an amino acid at the location of termination codon. Mutant tRNAs are known which are capable of recognizing termination codons and insert amino acids like *tyrosine, serine* or *glutamine* to restore the

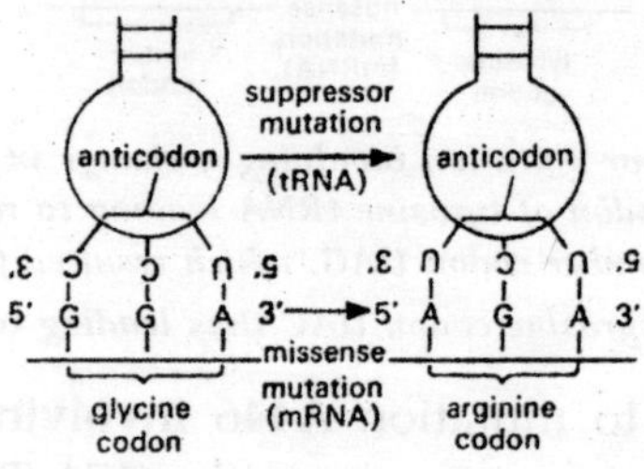

Suppressor mutation involving a change in one base of anticodon of glycine tRNA leading to misreading of arginine codon AGA (as glycine) which resulted due to missense mutation in glycine codon GGA thus leading to suppression.

function of protein (Fig.). However these mutant tRNAs may also read the normal stop signals of proteins thus causing elongation of protein beyond its normal termination thus causing disturbances. To prevent this disturbance, double stop signals are often available in mRNA, so that if mutant tRNA reads one, the other may still bring about termination. Moreover, two genes may code for same tRNA, so that if one (minor gene) mutates, and gives rise to suppressor tRNA the other (major gene) still functions and gives rise to normal tRNA.

Transfer RNA (tRNA) mediated UAG suppression was first observed, when in coat protein of phage R17, amber codon 5′UAG3' could be read by a tyrosine suppressor tRNA (Fig.), in which the anticodon had mutated from 5′GUA3' to 5′CUA3'. There are also other examples of this type of suppression.

Missense Suppression and Mutant tRNA : We have already discussed mutations resulting due to base substitutions leading to replacement of one amino acid by another. These are called missense mutations. These mutations will lead to synthesis of a protein with a wrong amino acid at a particular site. Such a mutation can also be suppressed like nonsense mutations, by mutant tRNA, which will misread the codon and thus substitute a correct amino acid, although the mutant codon codes for a different amino acid. For instance, in mutant A36 in *tryptophan synthetase* of *E. coli,* glycine (coded by GGA) is replaced by arginine (coded by AGA). A suppresser mutant may have tRNA, depositing glycine for codon AGA actually meant for arginine (Fig.).

Frameshift Mutations and Mutant tRNA : In later, a reference was made about frameshift mutations. Earlier in this chapter these mutations were discussed, where the reading frame of mRNA shifts to the right due to deletion or to the left due to addition of a base pair in DNA. It has been shown that a frameshift mutation due to a deletion can be suppressed by mutant tRNA having an anticodon of two bases rather than three (doublet anticodon rather than triplet). Similarly, addition frameshift mutations can be suppressed by mutant tRNAs each having a quadruplet anticodon instead of a triplet. Mutant tRNAs with doublet and quadruplet anticodons have actually been detected in biological systems.

Suppression and Ribosomal Mutations : Mutations in some ribosomal proteins may cause a change in ribosome structure, so that it may now misread the mRNA codons. One such mutation, called *ram* (ribosomal ambiguity) is capable of reading all the three termination codons, thus causing suppression of nonsense mutations. Some frameshift mutations may also be suppressed by these *ram* mutants.

Mutations may also occur in ribosomes rendering them sensitive to an antibiotic like streptomycin. These sensitive ribosomes may then allow misreading of mRNA codons, although in resistant ribosomes this misreading is much less frequent. For instance, when Poly-U is used as a template with sensitive ribosomes, most frequent error is replacement of phenylalanine (coded by UUU) by isoleucine (coded by AUU).

It has been shown that streptomycin sensitive strains carry mutations in 30S-ribosomal proteins, causing suppression through misreading.

'Second Genetic Code' and 'Second Half of the Genetic Code' : Earlier we have discussed *'Second Genetic Code'* and *'Second Half of the Genetic Code'* A distinction, therefore, needs to be made of these two expressions from *'The Genetic Code'* described in this chapter. As already mentioned earlier in this chapter, the genetic code is meant for translation of the language of nucleic acid into that of protein, using triplet codons. However, the 'Second Genetic Code' refers to the mechanism involved in the interaction of specific tRNAs with corresponding amino acyl synthetase enzymes (how does an enzyme recognize its tRNA or *vice versa,* and the 'Second Half of the Genetic Code' refers to the protein folding problem or to the rules, which govern the formation of a three dimensional structure from the primary structure (amino acid sequence) of a protein. Both these problems remain only partly resolved and continue to receive attention of molecular biologists.

X-ray crystallographic structures of several 'tRNA-synthetase complexes' representing intermediates in the formation of 'amino-acyl tRNAs' have been studied. The most important of these is a complex between 'glutamyl-tRNA synthetase' with $tRNA^{Gln}$ (tRNA for glutamine) and ATP. It was shown that there are multiple points of contact all along the inner side of the L-shaped structure of tRNA, which helps in recognition of tRNA by synthetase enzyme and in the formation of linkage between tRNA and amino acid.

Recoding of the Genetic Code

In recent years, examples have been discovered, where in a minority of cases, there is a set of instructions (called 'reading signals') contained in the mRNA itself, that specify an alteration in how the genetic code is to be applied. These alterations are described as 'recoding' and include (i) alteration in the linear read out mechanism, involving +1 or –1 shift (or a hop of many nucleotides) in the frame of reading, at a particular point, when mRNA is being read (meaning of codon words remains unaltered); (ii) alteration in the meaning of the code words.

Alteration in the Linear Read-out Frame : There are several examples of this recoding mechanism : (i) In *E. coli,* mRNA of RF2 (protein meant for termination of translation = release factor) programmes 30% of ribosomes to change to the + 1 reading frame after codon number 25 (codon 26 is UGA; see Fig.). The frameshift event is encouraged by an upstream sequence (termed 'stimulator') that pairs with 16S RNA in the ribosome. There are two components of recoding signals, a site of action (codons 25 and 26) and a stimulator, (ii) In a class of retroviruses, many mRNAs exhibit a ribosomal frame shift to the –1 reading frame by tandem slippage of tRNA. The two components of 'recoding signal' are, a site of action consisting of a heptanucleotide in the mRNA and a stimulatory sequence

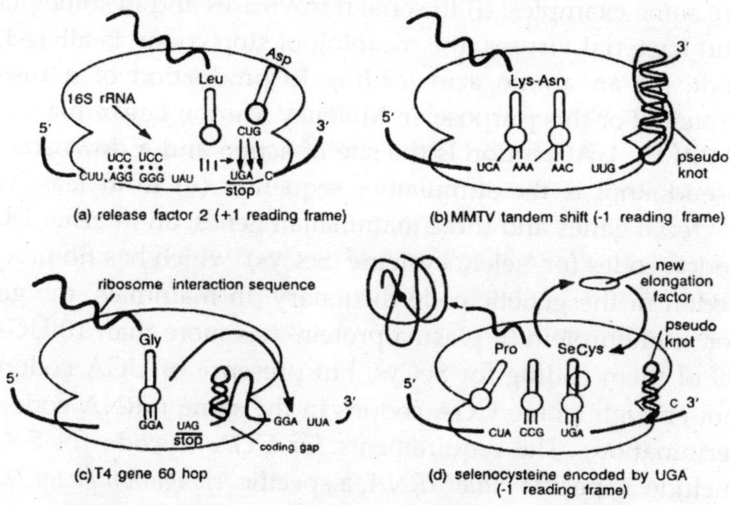

Different alternatives leading to recording of the genetic code

in the form of stem loop or pseudoknot structure (present downstream; see Fig.). A similar heptanucleotide for frameshift is found in *dnaX* gene of *E. coli,* where 50% of gene product in foreshortened due to frameshift. (iii) In higher animals, a protein called 'antizyme' renders the enzyme 'ornithine decarboxylase (ODC)' unstable. Decoding of antizyme mRNA requires a + 1 frameshift, which is regulated by the concentration of polyamines (product of ODC). (iv) In the mRNA for T4 *gene* 60, ribosomes hop a stretch of 50 nucleotides found between a pair of glycine codons. At the first glycine codon, 100% ribosomes are released; these ribosomes hop and read the second glycine codon downstream and continue translation of the rest of mRNA (Fig.). The stimulatory signals include the mRNA structure and the amino acid sequence of nascent peptide chain.

Alteration in the Meaning of Code Words : In almost all organisms, there seem to be available 'recoding signals' in mRNAs leading to altered meanings of codons. These are exceptions of the universal character of genetic code. Following are some examples: (i) In some retroviruses and in some plant and bacterial viruses, the meaning of stop codon is altered to code for an amino acid leading to production of a fusion protein. For this purpose, in Moloney Murine Leukemia Virus (MMLV), UAG codon is the site of action and a downstream pseudoknot is the stimulatory sequence. (ii) In at least two bacterial genes and three mammalian genes, an internal UGA codon codes for 'selenocysteine (SeCys)' which has no unique codon in the genetic code dictionary. In mammals, the gene for selenoprotein 'P plasma protein' has more than 10 UGAs, 10 of them coding for SeCys, but presence of UGA codon is not enough (some UGA codons in the same mRNA code for termination). The requirements for UGA to code for SeCys include a specific *minor tRNA,* a specific *'elongation factor (EF)'* and a particular *'downstream sequence'* (sometimes this downstream sequence is 200 nucleotides away in the 3' non-coding region; see Fig.).

2

Genetic Material

During the last two decades, the use of the techniques of molecular biology (including recombinant DNA technology) in genetic studies gave several surprises. Some of these surprises led to the discovery of the following : (i) split genes, meaning that the sequences containing actual information of the gene (exons) are interrupted by other sequences (introns) which are spliced out after transcription; (ii) overlapping genes, meaning that same DNA sequence can become part of each of two or more genes expressed at different times and in different reading frames; (iii) pseudogenes, which represent DNA sequences derived from mRNA through reverse transcription; these pseudogenes, therefore, differ from the split genes to which they belong, due to the absence of intron sequences; and (iv) cryptic genes, which are generally silent under a specific environment and may be activated (due to a variety of events) in a few individuals of a population in another environment; these genes differ from selfish DNA/selfish genes/ pseudogenes and offer a selective advantage to the organism, that carries them.

Split Genes or Interrupted Genes

Discovery and Nature of Split Genes : After rediscovery of Mendel's laws in 1900, a gene was considered earlier to be a point on a chromosome, and later to be a continuous segment of DNA. It was also proved as late as 1964 that there was co-linearity between sequence of nucleotides on DNA and sequence of amino acids in corresponding protein, so that there was absolutely no doubt about the continuity of nucleotides in a gene represented by a DNA segment. This was proved to be true not only in prokaryotes but also in eukaryotes. In view of this, it was a big surprise when, in 1977, it was proved for the first time that in some mammals, birds and amphibians, a gene may not be represented by a continuous sequence of nucleotides but may be interrupted by some intervening sequences which are not represented in mRNA transcribed from the gene and utilized for the synthesis of proteins. Such genes with intervening sequences were called split genes (Fig.).

The discovery of split genes was made in 1977 by several groups in a variety of materials, which included the following : (i) Two research groups separately headed by Phillip A. Sharp and Richard J. Robetrts studied genes of adenovirus 2. (ii) Research groups of D.S. Hogness, I.B. David and N. Davidson studied genes for 28S rRNA in *Drosophila.* (iii) Research groups of P. Chambon, P. Leader and R.A. Flavell studied β globin genes, ovalbumin genes and tRNA genes. In all these cases the genes were found to be interrupted by intervening sequences. The credit for discovery of split genes, however, goes to Phillip Sharp and Richard Roberts, who won the 1993 Nobel Prize for Medicine for their work on split genes. They analysed the hybrids of late mRNA of adenovirus 2 with the adenovirus genomic DNA. When these mRNA-DNA hybrids were examined under electron microscope, the adjoining sequences of mRNA were found to hybridized with discontinuous stretches of genomic DNA of adenovirus 2. The intervening DNA sequences were observed as loops and the phenomenon was later described as R-looping (see later for R-looping, see Fig.). Chambon's group compiled sequences of the boundaries of introns from a large number of protein coding eukaryotic genes (not ribosomal RNA or tRNA genes), which revealed the presence of consensus sequences at the intron-exon junctions. Of these GT was always found at the 5' side of the intron (left splice junction) and AG at the 3' side (right splice junction). This became popularly known as GT-AG or Chambon's rule (rarely, AT-AC occur at intron-exon junctions). Some of the diseases (e.g. thalassemians) are caused by mutations, which created or abolished these splice junctions.

A detailed study was conducted on *ovalbumin gene* found in chicken. This ovalbumin gene is responsible for synthesis of a protein 'ovalbumin' consisting of 386 amino acids and synthesized only by highly specialized tubular gland cells of the oviduct at the time when the hen is laying eggs. The expression of this *ovalbumin gene* is controlled by some female sex hormones. Pierre Chambon and his colleagues synthesized artificial ovalbumin gene in order to study its regulation. Such an artificial gene could be synthesized by using ovalbumin mRNA which could give rise to cDNA *(complementary DNA)* with the help of enzyme reverse transciptase. This cDNA was inserted into a plasmid and cloned in *E. coli* for its multiplication. When this cDNA was compared with corresponding genomic DNA, it was discovered (through DNA hybridization) that the genomic DNA had additional intervening sequences.

Another important technique utilized for the study of split genes was the use of restriction enzymes which have the property of cleaving DNA at unique sites. More than 100 such restriction endonuclease enzymes are now available. In view of this, restriction endonucleases could be used to find out the presence or absence of a certain sequence in a particular gene. For example, when *EcoR*1 and *HindIII* enzymes were used with cDNA for ovalbumin, it was found that no cleavage occurred suggesting that the sequences of six base pairs each recognized by these two enzymes were absent.

On the basis of this it was expected that if the DNA extracted from oviduct or another tissue was cleaved by utilizing these two enzymes, the *ovalbumin gene* will not be broken and the cleavage will occur at other places thus making it possible to isolate *ovalbumin gene* from the living cells. This DNA segment representing *ovalbumin gene* was expected to be separated with the help of hybridization with cDNA artificially synthesized. When such hybridization was done, it was found that cDNA could hybridize with different fragments of DNA rather than with a single fragment. Hybridization between the single stranded DNA having the gene for ovalbumin and its mRNA also showed formation of distinct loops at specific sites as observed in electron microscope (Fig.). This DNA forming the loops is obviously missing in mRNA.

This kind of results eventually led to the discovery of split genes in 1977. Subsequently it could be proved that split genes are present at least in two more cases i.e. gene for β-globin (a component of haemoglobin molecule) in rabbit and mouse and immunoglobulin gene (antibody gene). By the end of 1977,

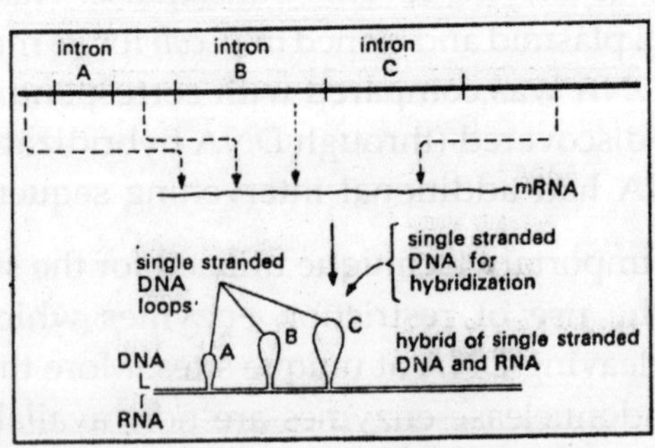

The upper figure shows a DNA sequence representing an interrupted gene with there introns (A, B, C), and the synthesis of its mRNA; the bottom figure shows the result of hybridization of mRNA with single stranded DNA obtained after denaturation of native DNA (note the loops formed by three intron regions).

it was fairly clear that the presence of split genes in case of higher organisms was a phenomenon of general occurrence. We may discuss the problem of split genes by taking a hypothetical example of a gene represented by a DNA sequence which has three pieces, called exon 1, exon 2 and exon 3 (Fig.). These three exons are separated by two long intervening sequences called introns A and B.

The terms exon and intron were used by W. Gilbert for the first time and are being followed ever since. Figure shows, that in the transcription of DNA, the ultimate product *i.e.* mRNA had only those sequences which corresponded to exons, the sequences representing introns being entirely absent. The mechanism of production of mRNA from DNA sequences having intervening sequences called introns was a matter of discussion during 1978 and 1979. Later it was discovered that both exons and introns are first transcribed and that this primary transcript is then modified.

The sequences corresponding to introns are removed and the sequences corresponding to exons are joined together, in correct order to give rise to mRNA. A generalisation has also been made that the order of exons on DNA is the same as the order in which they are found in processed mRNA. There does not seem any strong reason why it should always be true and it is possible that the order of exons would sometimes be different.

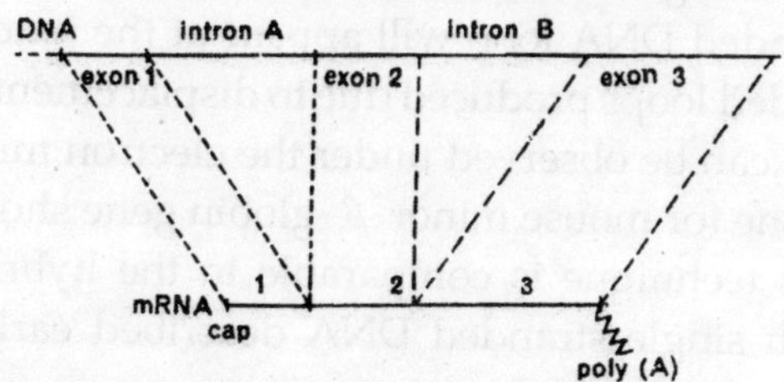

Diagrammatic representation of a hypothetical split gene having three exons (exon 1, exon 2, exon 3) and two introns (intron A, intron B) and its relationship with mRNA. There are no sequences in mRNA corresponding to those in two introns (redrawn from Science. Vol. 204, 1979).

It has also been shown that all classes of genes can be interrupted. These interrupted genes may include :

(i) nuclear genes for proteins,
(ii) nuclear genes coding for rRNA,
(iii) nuclear genes coding for tRNA,
(iv) mitochondrial genes in yeast,
(v) chloroplast genes in a wide variety of plants,
(vi) genes in archaebacteria and
(vii) genes in bacteriophages of *E. coli :*

They were initially believed to be entirely absent in eubacterial genomes, but during 1990s, introns were discovered even in eubacteria.

R-loop Mapping and Restriction Mapping of Interrupted Genes : R-loop mapping is a technique in which mRNA is hybridized with double stranded DNA under conditions in which RNA-DNA hybrid is more stable than the DNA duplex. This allows displacement of one strand of DNA in the region, where it can pair with RNA. When DNA segment belonging to an interrupted gene is used in such an experiment for hybridization with mRNA, the intervening sequence or intron region loops out as double stranded DNA and the exon region loops out as single stranded DNA as shown in Figure. The double stranded DNA loop will appear at the junction of two single stranded loops produced due to displacement by mRNA. These loops can be observed under the electron microscope as has been done for mouse minor β-globin gene showing single intron. This technique is comparable to the hybridization of mRNA with single stranded DNA described earlier.

An interrupted gene can also be studied by comparing the restriction map of the interrupted gene with that of its corresponding cDNA prepared from its mRNA through reverse transcription. When this is done, there may be two situations :

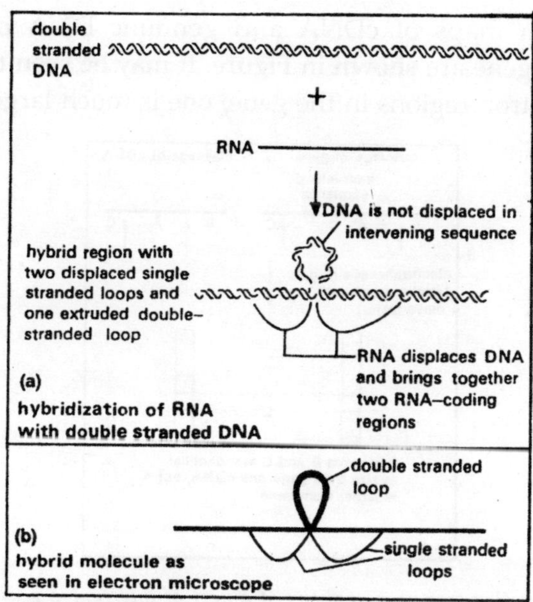

R-looping, where mRNA is hybridized with double stranded DNA (note the double stranded thick loop showing intron region and single stranded thin loops representing the exons).

(i) If the intervening sequence or intron region does not have a recognition site for an enzyme, there is only a change in the length of corresponding segment (since the two restriction sites in the regions on either side of an intron are now separated by a longer distance) but the number of segments remains the same (Fig.). (ii) If the intron region has a restriction site for the enzyme employed, an extra cut will be made in this region. This extra cut will produce two fragments corresponding to one segment in the cDNA (Fig.).

When restriction maps are prepared for cDNA and the gene using a number of enzymes, we may find that the two maps may be similar at the two ends but extra sites may be present in the intron regions of the gene, which are absent in cDNA or mRNA. The resolving power of restriction mapping allows detection of gene segments up to 20-30 bp long.

Restriction maps of cDNA and genomic DNA of mouse β-globin gene are shown in Figure. It may be seen that there are two intron regions in the gene; one is much larger, which

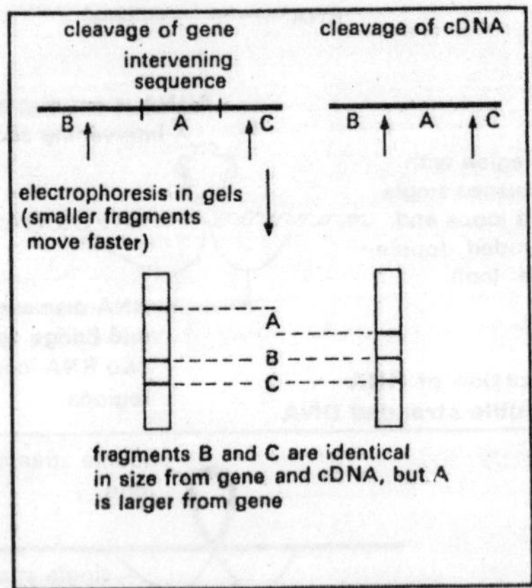

Restriction fragment analysis of a part of interrupted gene and its cDNA, showing that the length of a segment A (which had no restriction site) is decreased in cDNA due to splicing out of intron region.

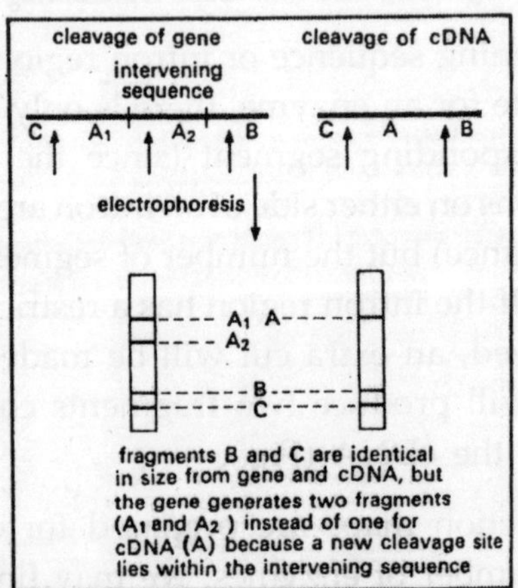

Restriction fragment analyses of a part of interrupted gene (having a restriction site in the intron region A, dividing it into two regions A_1 and A2) and its cDNA showing decrease in the number of fragments in cDNA from four to three.

can be visualized in R-loops under the electron microscope, but the other is smaller, that can not be visualized under the electron microscope. This demonstrates the higher resolving power of restriction mapping.

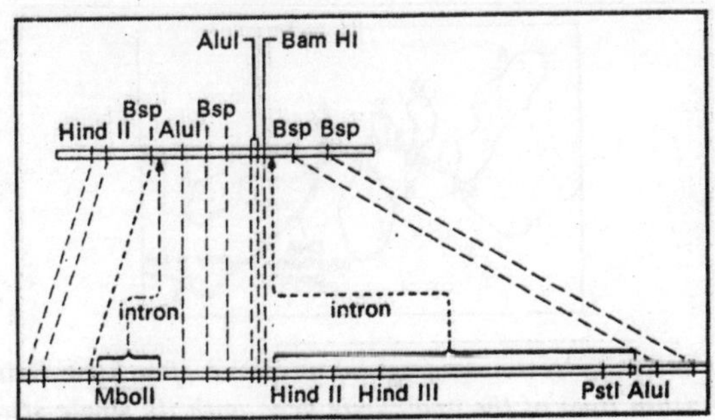

A comparison of the restriction maps of cDNA and genomic DNA for mouse β -globin gene, showing that the gene has two introns not present in cDNA. Only one of them shows up as a loop during hybridization of mRNA with denatured single stranded DNA (since the second intron is very small).

Structure of Chicken Ovalbumin Split Gene : The structure of ovalbumin gene in chicken has been studied in considerable detail and it has been found that in the split gene of ovalbumin there are eight exons and seven introns (Figs.). The size of ovalbumin gene has been shown to be 7,700 base pairs although the mature mRNA for ovalbumin is only 1,872 nucleotides. It has also been shown, that entire ovalbumin gene having 7,700 base pairs is first transcribed to a precursor RNA to which a cap consisting of 7mG is added at the 5' end and poly A tail is added at 3' end.

After the addition of cap and tail, the introns are excised, perhaps in two installments, five introns being excised in the first step and the remaining two in second step. The exons produced due to splicing are then joined with the help of enzyme ligase to produce the mature mRNA of *ovalbumin gene.* The different steps involved in the processing of precursor RNA into mRNA for *ovalbumin gene* are shown in the Figure.

Split Genes in Fungal Mitochondria : Split genes are also found in mitochondria. Introns of these split genes in fungal mitochondria are classified into group I and group II introns on the basis of their internal organization, (i) Group I introns,

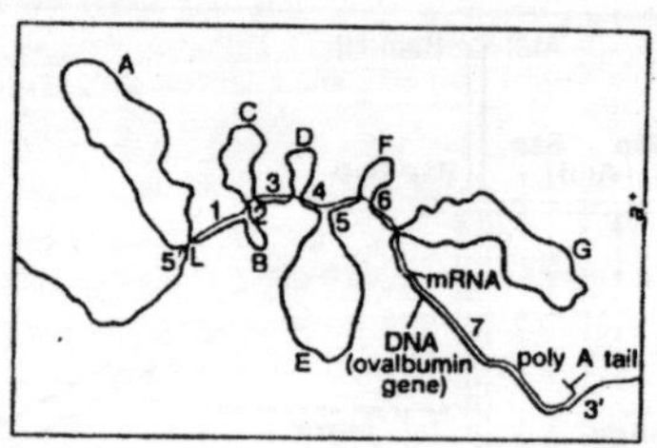

The map of electron micrograph prepared after hybridizing mRNA (thin line) of the ovalbumin gene with its single stranded DNA (thick line). The loops A-G represent 7 different introns and segments L and 1-7 represent 8 exons.

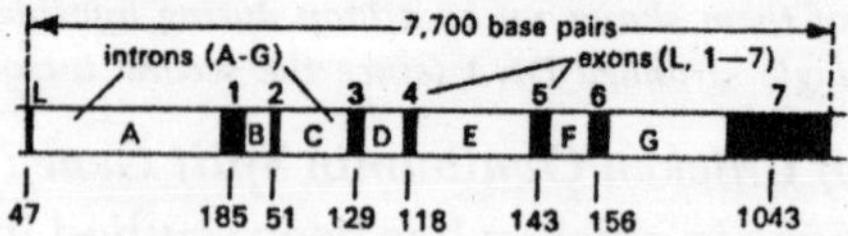

Schematic representation of ovalbumin gene showing 7 introns (A-G) and 8 exons (L, 1-7); number of base pairs in each exon are also given.

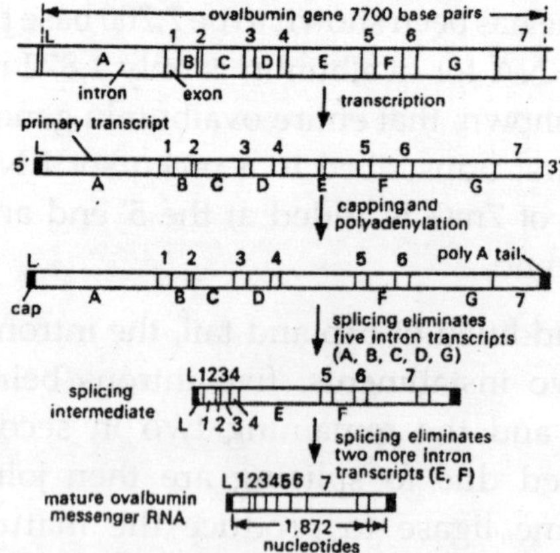

Steps involved in the production of ovalbumin mRNA (1872 nucleotides long) from ovalbumin gene (7700 base pairs long); introns (A-G) and exons (L, 1-7).

which are found in majority of the fungal mitochondrial split genes, do not carry any conserved sequences at intron-exon junctions, but carry internally a short conserved sequence called 'internal guide sequence'. By internal pairing, this guide sequence brings the two intron-exon junctions together and help in splicing out the introns. These group I introns are also found in the nuclear genes coding for rRNA in *Tetrahymena* (a ciliate) and *Physarum* (a slime mold). Features similar to those of group I introns are also found in introns of phage T4 genes, (ii) Group II introns resemble nuclear genes and have consensus sequences (GT and APy) and a branch sequence that resembles the TACTAAC box. These introns are excised as lariats.

Split Genes in Chloroplasts : Split genes for ribosomal RNA (rRNA), transfer RNAs and some proteins have also been reported in the chloroplast genomes of several plants including *Chlamydomonas* and *Nicotiana*. Introns found in chloroplast genes can be classified into three groups on the basis of intron boundary sequences. (i) Group I introns (e.g. in *trnL)* can be folded in a secondary structure similar to self splicing rRNA precursor of *Tetrahymena*. These can be removed either by self splicing or by a 'maturase' (as in cytochrome b and cytochrome oxidase mRNA precursors). (ii) Group II introns (e.g. majority of genes including *trnA* and *trnl)* can be folded into a complex secondary structure (as in introns of mitochondrial genes for cytochrome oxidase in maize and yeast). (iii) Group III introns (e.g. *trnG, trnK, trnV, rpl2, rps12, rps16,* etc.) have conserved sequences at their borders (GTGCGNY at 5' end, and ATCNRYY(N)YYAY at 3' end), similar to those in the eukaryotic nuclear genes (R = purine; Y = pyrimidine; N = any nucleoide).

Base Composition and Length of Introns vs Exons in Plants: A comparison of base composition between introns and exons of genes in plants revealed that plant introns are richer in AT base pairs than their adjacent exons. In dicots, the AT base pairs range from 59-83% with a mean of 74% as against 55% AT base pairs in their adjacent exons. Similarly in monocots, it rangs from 31-77% with a mean of 59% as against 44% AT in their adjacent exons. In addition, plant introns are short, the average size being 80-139 nucleotides (two third of all introns studied are less than 150 nucleotides).

In both these features (base composition and length), plant introns differ from vertebrate and yeast introns, although they resemble introns in animals, like fruitfly *(Drosophila)* and the extensively studied nematode, *Coenorhabditis elegans.* These key features of plant introns provide them with specificity during pre-mRNA splicing of these introns. This splicing is not so sensitive to the environment (temperature, etc.) as that of budding yeast, fruitfly and mammalian cells. In other words, the plant introns are spliced accurately over a broad temperature range, while animal introns are sensitive to environment fluctuations.

'Intron enhancement' gene expression in transgenic cells. It has been shown,.that addition of specific intron sequences to gene constructs used for the production of transgenic plants, leads to high level of expression of the associated genes. This phenomenon is described as 'intron enhancement' and the extent of enhancement in expression depends on (i) plant system (monocot or dicot), (ii) expression system (transient or transgenic), (iii) the intron and the promoter used and (iv) the site of insertion. The most commonly used intron for enhancement of the expression of transgene is maize *adh1* intron 1, which increases the expression of (i) CAT mRNA from an *adh1* promoter, 16-fold and (iii) NPT II mRNA expression from CaMV35S promoter, 15-fold. There are many other examples of intron enhancement achieved in transgenic crops by using the above *adhl* intron. (CAT, luciferase and NPT II are products of reporter genes used for the production of transgenic plants).

Intron of One Gene May Contain Exon of Another Gene : Sometimes more than one mRNA sequences may be derived from the same DNA segment, due to tissue specific expressions in different tissues. In this process, part of an intron for one gene is used as exon for another gene. Following four examples illustrate this phenomenon. (i) Amylase gene in mouse has two promoters, one of them expressed in liver and the other expressed in salivary gland. Consequently the 5' ends of precursor mRNAs differ. This leads to the presence of different coding sequences as exon 1 in the two cases (*exon S* in salivary gland and *exon L* in the liver). The derivation of two mRNAs from the same gene encoding two amylases is shown in Figure. It may be noted that exon L is actually derived from a part of

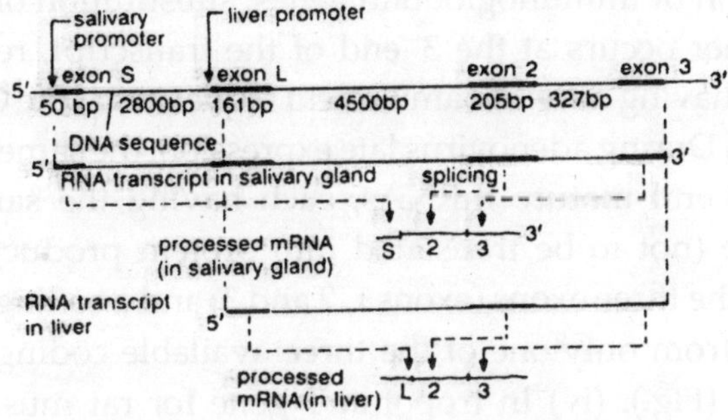

In mouse, derivation of two different amylase mRNAs from same DNA sequence, due to the presence of two promoters and differential RNA splicing.

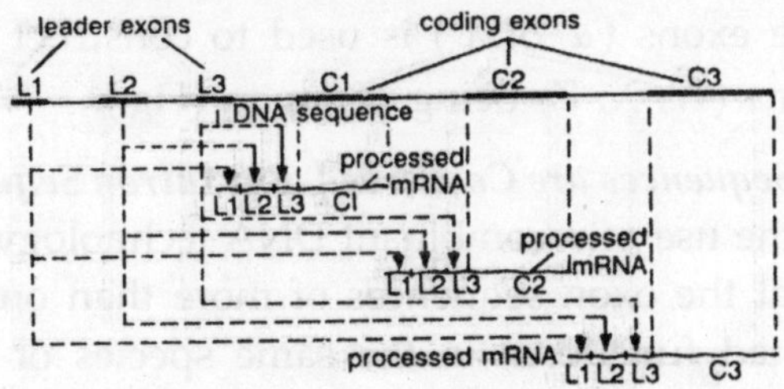

During adenovirus late expression, derivation of different mRNAs from same DNA sequence, due to cleavage, so that same leader sequence (three exons) is joined with different exon sequences.

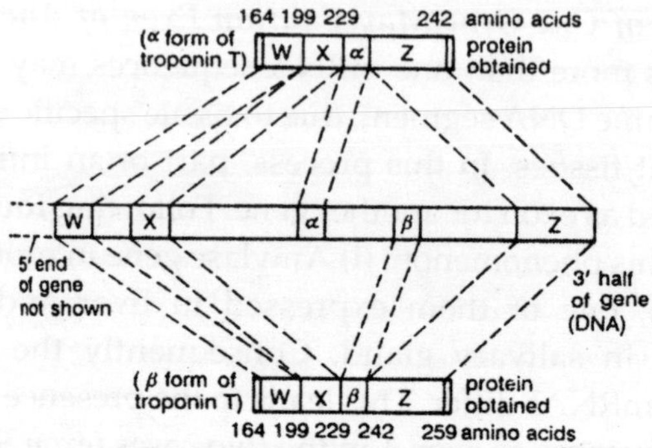

Derivation of α and β forms of troponin T from same DNA sequence, due to differential splicing patterns in rat muscle.

the intron of the amylase gene for salivary glands. (ii) During expression of immunoglobulin genes, substitution of one exon by another occurs at the 3' end of the transcript, resulting in proteins having different amino acid sequences at the C-terminal ends. (iii) During adenovirus late expression, the same transcript gives several mature mRNAs, each having the same leader sequence (not to be translated into protein product) derived from all the three exons (exons 1, 2 and 3) and a coding sequence derived from only one of the three available coding exons A, B and C (Fig.). (iv) In troponin T gene for rat muscle, at the 3' end there are five exons, of which only four each take part in forming mature mRNAs for two different forms of troponin T. The two forms differ, depending upon which of the two alternative exons (α or β) is used to construct mRNA, the other exons (W, X, Z) being common (Fig.).

Exon Sequences are Conserved, But Intron Sequences Vary : Through the use of recombinant DNA technology, it could be shown that the exon sequences of more than one genes (for some related functions) in the same species or in different species, often have similar DNA sequences. However, the corresponding intron sequences differ substantially. This was demonstrated for two β globin genes of mouse. This property has been utilized in selecting exon sequences of heterologous genes for isolation of related genes in different species.

Introns with Coding Sequences : In fungal mitochondria, for some genes, it was shown that introns may code for RNA maturases or endonucleases which take part in RNA processing, and DNA recombination respectively.

Two Hypotheses for Origin of Introns 'Early Intron' and 'Late Intron' : Although introns are generally absent in protein-coding genes of bacteria, archaebacteria and some lower eukaryotes, class II type of self splicing introns have recently been reported even in bacteria. Such a distribution of introns has been used for a study of the role of introns in the evolution of genes or alternatively for a study of the origin of introns themselves. There are two different hypotheses available in this connection. (i) According to 'early-intron hypothesis' (also called 'exon theory of genes' by Walter Gilbert in 1987), ancestral exons existed as independent genetic units or microgenes and their association with introns helped generation of diversity in genes through trans-splicing and exon shuffling. There are examples, where new genes could be generated by exon shuffling through intron recombination events. This and several other known events are used as evidences in favour of early intron hypothesis, (ii) According to 'late-intron hypothesis' (also called the 'insertional theory of intron origin') intron evolution was independent of the evolution of protein-coding genes. It is assumed in this hypothesis, that the self-splicing introns invaded the genome of eukaryotic ancestor, as endosymbionts and evolved into mitochondria and chloroplasts containing class II introns.

It is postulated that invasion took place at the mRNA level, which was followed by reverse transcription and recombination with DNA, producing split genes. These introns gave other types of introns later. This explains, why introns are absent in bacteria, archaebacteria and lower eukaryotes. The recent report of introns in bacteria may be the initial stage of this process, of the evolution of introns. Efforts are being made to combine the two above hypotheses using strong features of both to formulate a new intermediate hypothesis.

Overlapping Genes

We will later discuss the genetic code, which has been shown to be non-overlapping and triplet. In view of this if the length of DNA is known, then the maximum length of protein, which can be synthesized, can be worked out in terms of number of amino acid residues. For instance, a DNA segment of 1200 base pairs can give rise to a protein having not more than 400 amino acids.

Moreover, if number of amino acids in proteins synthesized, and number of nucleotide base pairs in DNA are known, they can be compared and one would expect that the number of base pairs should be at least 3 times the number of amino acids in the protein. Rather if there are segments of DNA which are not coding for protein, then the length of DNA segment will be more than three times the number of amino acid residues in protein.

Such comparisons have been made in bacteriophage ϕX174 which contains a single stranded DNA approximately 5,400 nucleotides in length. Complete sequence of these 5,400 nucleotides was also worked out by Fredrick Sanger and his colleagues at the MRC Laboratory of Microbiology, Cambridge, U.K. (Fredrick Sanger for this and other work got the second Nobel Prize in the year 1980).

The genome of ϕX174 consists of nine cistrons. Amino acid sequences of major proteins coded by the genome of (ϕX174 have also been worked out) and their molecular weights are known. From the information about proteins coded, an estimate could be made of the number of nucleotides required. This estimate of number of nucleotides exceeds 6,000 which is much higher than the actual number of nucleotides present i.e., 5,400. Therefore, it was difficult to explain how these proteins could be synthesized from a DNA segment which is not long enough to code for the required number of amino acids. On detailed study of the system, it was discovered that sequences in the same segment could be utilized by two different cistrons coding for different proteins. Such overlapping of cistrons will be theoretically possible if the two cistrons have two functions at different times and their nucleotide sequences are translated in two different reading frames.

In 1976, Barrell and his co-workers discovered that in ϕX174, having nine cistrons (A, B, C, D, E, J, F, G, H), cistron E is present between D and J and that the cistron E overlaps cistron D. This could be done by working out the nucleotide sequences of DNA segment representing gene D, E, and J. It could be shown that amber mutations in cistron E lie within the cistron D and, these amber mutations do not influence the translation of cistron D into its protein.

Similarly some nonsense mutations for cistron D lie in cistron E suggesting that the cistrons D and E overlap in the DNA sequences and that the cistrons D and E are translated in two different reading frames so that amber codon in mRNA of one cistron will not be read as termination codon during the translation of mRNA of the other cistron.

Among genes D, E, and J in ϕX174, cistron D codes for a simple protein, cistron E codes for a protein for lysis and cistron J codes for a particle component. It was shown that enough space is not available where these three genes are located, thus confirming that there was overlapping between D and E cistrons. Overlapping has also been suggested to be present in regions of genes A and B and mutants were located in overlapping zone. Overlapping regions were also found to be present at beginnings of cistrons A, C, and J, where in each

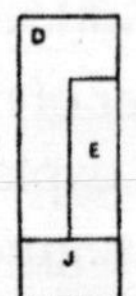

***Overlapping of E and D genes in phage* ϕX174 .**

case the initiation codon of a cistron overlaps with the termination codon of its corresponding previous cistron. A DNA segment showing overlapping of D and E is shown in Figure.

Overlapping sequences have also been discovered in tryptophan mRNA of *E. coli* and several other cases. In view of wide occurrence of overlapping genes, it seems that phenomenon of overlapping genes, is an economic device to make better use of genetic material, through packing of more genetic information in less DNA. This phenomenon may be of general occurrence. Therefore, it may be necessary to modify our concept of coding potential (number of amino acids, which a DNA segment can code for) of DNA genomes, because a DNA can potentially code for many more amino acids in different reading frames than the number of amino acids expected to be coded, if no overlapping was present.

Pseudogenes

(U snRNA Series, Globin Pseudogenes and Alu Family) : In multicellular organisms, a wide variety of DNA sequences are found, which are of no apparent use. Some of these sequences are defective copies of functional genes and are, therefore, called pseudogenes. These pseudogenes have been reported in human beings, mouse and *Drosophila.* The most popular examples of these pseudogenes include the following, (i) Human α-globin and β-globin pseudogenes (ψ), found in each of the two globin gene clusters. Complete nucleotide sequence of alpha globin, psudogene is now known and it has been shown that both these genes are non-translatable, since they may have mutations in initiation codon and also frame-shift mutations along their length. (ii) In mouse also there are two alpha globin pseudogenes (ψ), one of them ($\psi\alpha3$) is different from other pseudogenes since it has no introns which are present in functional α-globin genes as well as in other pseudogenes. The pseudogene $\psi\alpha3$ is, therefore, believed to be the reverse transcript of mRNA, obtained due to reverse transcriptase enzyme. (iii) U snRNA series of pseudogenes in human beings includes U_1, U_2 and U_3 snRNA pseudogenes. The corresponding functional U_1 U_2 and U_3 snRNA genes may be involved in splicing of nucleic acid sequences. (iv) In *Drosophila*, histone pseudogenes have been discovered. (v) There are other sequences in human genome which are popularly called Alu family of dispersed repetitive elements, of which there are about 300,000 copies in a haploid human genome. Each sequence is 300 bases long and they are capable of being transcribed by RNA polymerase III (meant for 5SRNA and tRNA), rather than RNA polymerase II (meant for mRNA).

There is a number of characteristic features of these pseudogenes which throw light on their possible origin. These characteristics include the following : (i) Most pseudogenes outnumber their genes and are therefore repetitive sequences. (ii) Most pseudogenes are flanked by short direct repeats, — 6–21 bases long, suggesting their subsequent insertion.

Such pseudogenes are believed to have originated by one of the following methods. *First* possibility is that the mRNA or snRNA itself becomes incorporated into retrovirus RNA, followed by the incorporation of retrovirus into the genome; *Second* possibility is that reverse transcript of mRNA or snRNA was synthesized as cDNA, which subsequently became integrated. It is this second alternative which is considered to be the more probable mechanism of the origin of pseudogenes. In this hypothesis RNA takes up a configuration favourable for synthesis of DNA at its 3′ end copying a part of this RNA as a template to form cDNA. The insertion of this cDNA due to staggered cleavage of genomic DNA at a specific site is shown in Figure.

Cryptic Genes

Cryptic genes are defined as phenotypically silent DNA sequences that are not expressed during the normal life cycle

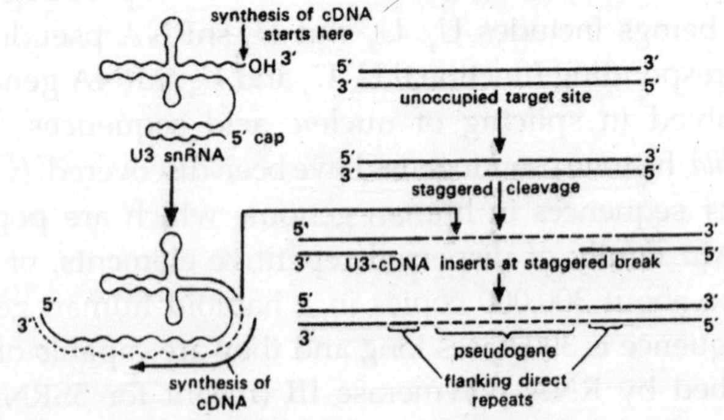

A model for origin of pseudogenes, in which cDNA (complementary DNA) is synthesized on snRNA (small nuclear RNA) and is inserted in DNA, which undergoes staggered cleavage through restriction endonuclease activity.

of an organism. They may be activated in a few individuals of a population, due to events like mutation, recombination, transposition or other genetic mechanisms (or due to specific altered environmental conditions), and this leads to their detection.

They, however, differ from other silent DNA sequences, collectively referred to as 'selfish DNA', which spread in the genome by making additional copies of themselves, without conferring any phenotypic advantage. They also differ from pseudogenes, which are homologous copies (often derived from mRNA due to reverse transcription) of active genes, but carry many mutations, so that their reversion to active form is not possible.

There are many examples of cryptic genes including the following:

(i) Genes for β-glucoside utilization in *E.coli* and *Solmonella.* Four cryptic genetic systems *(bgl, cel, asc, arbT)* involved in the β-glucoside utilization have been characterized in *E.coli.* They mutate spontaneously to a phenotype detectable as papillae on the indicator plates when supplemented with specific β-glucoside.

(ii) Gene for citrate utilization in *E.coli.*

(iii) A gene for alcohol dehydrogenase *(Adh* IV) in yeast.

These cryptic genes are characterized by the following two features: (i) The cryptic state is a transient state and oscillations between cryptic and active states are directed by changing environments, which impose different fitness values to the two states. In a selective environment the active form prevails in the population (ii) Cryptic genes are silent only under the conditions in which they are observed and may become functional even in the absence of an activating mutation under specific environmental conditions.

In view of this, it appears that cryptic genes confer selective advantage to the organisms that carry them. This explains the continued presence of these cryptic genes in the genome of an organism.

Promiscuous DNA

(Movement of DNA Between Mitochondria, Chloroplasts and Nucleus) : During 1982-84, several evidences suggesting transfer of DNA from chloroplasts to mitochondria and from both these organelles to the nuclear genome, have been reported. For instance, it was suggested that apparent translocation of DNA sequences from chloroplasts to mitochondria has taken place in corn. This conclusion was based on the discovery that a 12 kilobase segment of mitochondrial DNA (mtDNA) was homologous to a section of chloroplast DNA (cpDNA) of same organism, corn. During the same period, a report from Texas, U.S.A., described the same phenomenon in yeast, where a segment of mtDNA showed similarity with a segment in nuclear DNA, suggesting flow of DNA from mitochondria to nucleus. A similar report of transfer of mtDNA to nucleus in sea urchin also came from California Institute of Technology.

In 1984, more reports of promiscuous DNA came from different places in materials like mungbean, spinach, corn and peas, suggesting that movement of DNA sequences among organelles and between organelles and nucleus must be very common and not exceptional. The movement of DNA among mitochondria, chloroplasts and nucleus are shown in Figure. The evolutionary implications of promiscuous DNA are far reaching and it is speculated that chloroplasts and mitochondria, before invading eukaryotic cell were free living primitive prokaryotes.

The endosymboint hypothesis suggests that in course of time these entrapped organisms lost their independence by losing genetic information to nuclear genome. Thus the promiscuity of DNA gave further support to endosymboint hypothesis, but exchange of DNA between mitochondria and chloroplasts is surprising, since they are independent of each other. This may be partly explained on the basis of great variation and bigger size of mtDNA, relative to cpDNA which is tightly constrained in size and sequence even between species.

While mtDNA can tolerate foreign DNA and other disturbances, chloroplasts are sensitive to any such variation or modification. This is confirmed by evidence produced at Carnegie Institute of Washington, Stanford, U.S.A., where it was shown that chloroplast sequences from a variety of plants are scattered throughout mitochondrial genomes, and some of these transfers are believed to be rather recent.

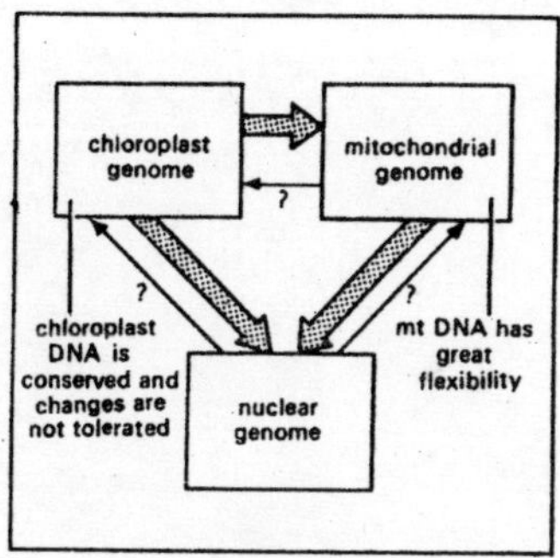

Exchange of genetic material between nuclear genome, chloroplast genome and mitochondrial genome. Thick arrows indicate transfer of promiscuous DNA for which evidence is available.

The mechanism of this transfer of so called promiscuous DNA is not known but two possibilities are considered. (i) Some kind of a vector or transposon or transducing phage may be involved. (ii) The two organelles might have undergone fusion at some stage. This second mechanism involving fusion is preferred on the basis of electron microscopic observations.

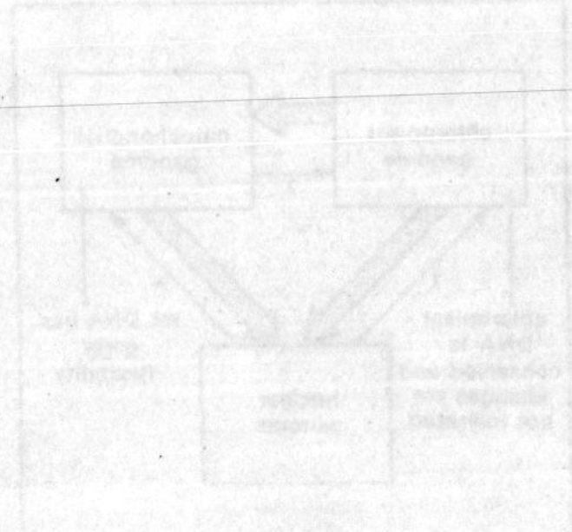

3

Organization of Genetic Material

During last three decades, extensive survey has also been conducted for the DNA contents per genome in prokaryotes and eukaryotes.

Large variation in this DNA content has been witnessed particularly in eukaryotes, where the quantity of DNA per cell is enormous, when compared with that in prokaryotes.

Large proportion of this DNA in most eukaryotes has been shown to be present in the form of multiple copies of identical DNA sequences, this multiplicity ranging from a few copies to millions of copies within the same cell. These DNA sequences are together described as repetitive DNA.

The remaining DNA in the cell is found in the form of single copy DNA sequences and is therefore described as unique DNA.

While the function of repetitive DNA is largely unknown,

the unique DNA contains most of the genetic information in the form of structural genes controlling phenotypic traits directly or indirectly.

The repetitive DNA may be clustered at one place or interspersed with unique sequences, each repeat unit ranging from a few hundred to few thousand base pairs.

These aspects of organization of DNA in the form of repetitive and unique sequences will be discussed in this chapter.

Chromosomal DNA Content and C-Value Paradox : Large variation in DNA content, among species of the same genus as well as among genera of the same family have been observed. This may be partly attributed to differences in chromosome numbers.

However, species at the same ploidy level and with same chromosome number may either show little or no variation (e.g. *Hordeum, Avena,* etc.) or may exhibit many fold differences (e.g. *Lathyrus, Vicia, Helianthus, Crepis, Allium,* etc.).

This interspecific variation may be continuous or discontinuous. Intraspecific variation in DNA content has also been reported in recent years, so that the variation of DNA among plants and animals is a rule rather than an exception.

This variation has sometimes also been used to explain the mechanism of evolution in specific groups.

In Table and also in Figure, contents of DNA in some plants and animals are presented, which clearly indicate that eukaryotes, in comparison to prokaryotes, have enormous DNA per cell.

For instance, each human diploid cell has DNA equal to 3.5×10^{12} daltons (1 dalton = mass of one hydrogen atom),

DNA Contents in Some Organisms Showing Large Variation
(1 picogram = 6.1×10^{11} daltons = 965×10^6 base pairs)

Class of organism	Species	Number of genes	Haploid DNA (1C)		
			Picorgam	Daltons	Base pairs
1. Phages	$\phi \times 174$ (single strand)	11	2.8×10^{-6}	1.7×10^6	5,387
	Lambda (λ)	~ 40	46.7×10^{-6}	32×10^6	48.502
	T_4	~ 200	20.7×10^{-5}	126×10^6	165,000
2. Animal virus	Adenovirus	~ 50	21.7×10^{-6}	13 × 106	21,000
3. Prokaryotes	*Bacillus subtilis*	~ 2000	2.18×10^{-3}	1.3×10^9	2.1×10^6
Escherichia coli		-2000	4.4 × 10–3	2.7×10^9	4.2×10^6
4. Unicellular eukaryotes	*Saccharomyces cerevisiae*		14×10^{-3}	8.5×10^9	1.4×10^7
5. Multicellular eukaryotes	*Drosophila melanogaster*	~ 5000	0.18	0.11×10^{12}	0.17×10^9
	Zea may (corn) < 25,000	2.6	1.6×10^{12}	2.5×10^9	
	Xennpus laevis (toad) < 25,000	3.2	1.9×10^{12}	3.1×10^9	
	Mus musculus (mouse) < 25,000	2.7	1.7×10^{12}	2.6×10^9	
	Homo sapiens (human) < 25,000	2.8	1.8×10^{12}	2.8×10^9	

which is equal to 174 cm of DNA. Since a human body may have up to as much as l00g of DNA in all cells, this DNA in the body of a human will be as long as 2.5×10^{10} km (more than 100 times the distance from the earth to the sun). If the haploid genome content in humans is 1.8×10^{12} daltons, which is equal to 2.8×10^{9} base pairs, then this genome should contain approximately as many as 3×10^{6} genes (assuming 1000 base pairs per gene coding for nearly 300 amino acids.

However, an estimate of number of genes in humans has also been made on the basis of knowledge about the rate of mutation per locus. If average mutation rate per locus is known, and the mutation rate per individual is also known, then total number of genes can be calculated. The frequency of deleterious mutations per locus in humans is 10^{-5} to 10^{-6} in each generation as estimated by Muller in 1967. If number of genes is 3×10^{6} (3 million = 3,000,000) as outlined above, then it will yield 30 deleterious mutations in each generation at the rate of 10^{-5} mutations per locus. This will be an unbearable genetic load. However, the actual frequency of deleterious mutations per generation per individual has been estimated to be 0.5 against an expected frequency of 30. This mutation frequency at the rate of 10^{-5} mutation per locus will give an estimate of 5×10^{-4} genes (50,000 genes). This suggests that actual number of genes in a human should be 50,000, and not 3 million (3,000,000) as estimated from DNA content. This anomalous situation has been described by some workers as C value paradox (the haploid DNA content in an individual is described as its C value).

There is yet another aspect of C-value paradox involving huge variation in C-values between species, whose apparent complexity does not vary correspondingly. For instance, animals usually contain 1-5 pg DNA per haploid genome, but the range of this DNA in salamanders extends up to 100 pg or so and in primitive angiosperms it may extend even up to 300 pg.

Substantial variation in the nuclear DNA content exists even between closely related species (Fig.) and can not be easily explained and hence the phrase, *C-value paradox.*

Repetitive DNA

Technique for Detecting Repetitive DNA : In the year 1964, it was discovered with surprise that much of DNA in mouse cells consisted of multiple copies of same or, very similar sequences of base pairs. It is now known that in higher organisms, this repetitive DNA generally constitutes at least 20% and can reach up to as much as more than 90% in some cases. It is believed that these repetitive sequences do not carry any genetic information and, therefore, do not form genes, but play some other structural or regulatory role. The presence of

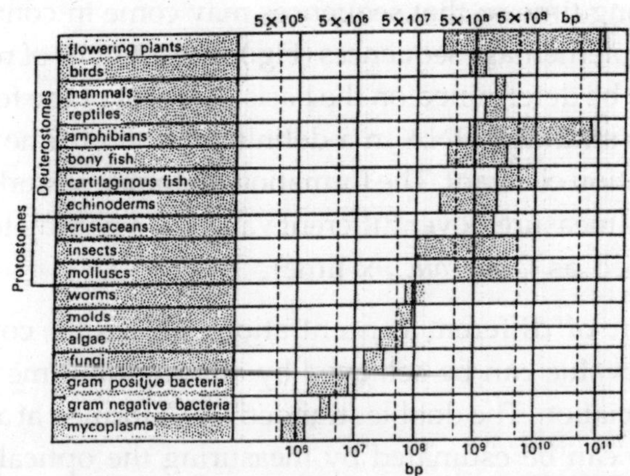

Variation in genomic DNA in a variety of animals and plants

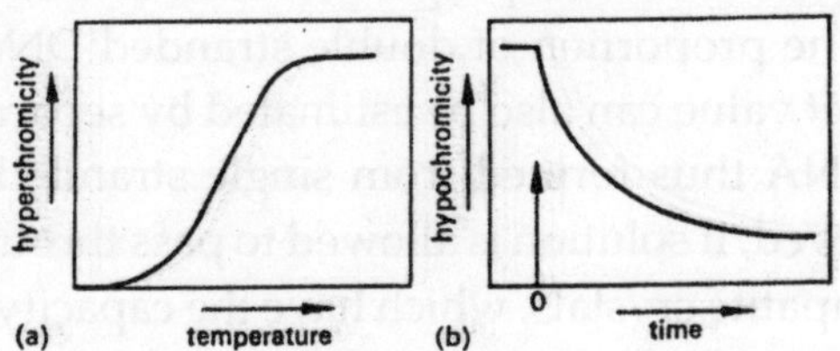

Hyperchromicity and hypochromicity of DNA due to denaturation and reassociation respectively.

repetitive DNA can be detected by a simple method involving following procedure : The double stranded DNA is first denatured and converted into single stranded DNA by heating the DNA solution. This is accompanied with increase in optical density, a phenomenon described as hyperchromicity. The solution of single stranded DNA thus obtained is then allowed to cool slowly, so that the double stranded DNA will be formed again, which accompanies decrease in optical density (hypocromicity, Fig.). This process is called reassociation of DNA for which one should maintain fairly high concentration of DNA and a temperature 25°C below the dissociation temperature or melting temperature Tm (Tm is a temperature, at which 50% reassociation is achieved).

The mixture should be allowed to undergo reassociation for a fairly long time so that sequences may come in contact with their complementary sequences (Fig.). The amount of repetitive DNA will be determined on the basis of extent of the formation of double stranded DNA in a definite period of time keeping concentration constant. The formation of double stranded DNA is actually measured over different values of a parameter which is described as Cot *(conc.,* x time).

If solution of different concentrations are to be compared, same *Cot* value can be achieved by altering the time allowed for reassociation. The double stranded DNA formed at a definite *Cot* value can be estimated by measuring the optical density of solution in a spectrophotometer, because the density of a solution goes down as the proportion of double stranded DNA increases. The proportion of double stranded DNA formed at a definite *Cot* value can also be estimated by separating double stranded DNA thus formed from single stranded DNA. This can be achieved, if solution is allowed to pass through a column of hydroxyapatite crystals, which have the capacity of retaining double stranded DNA and permitting single stranded DNA to pass through (Fig.).

If a DNA sample has high proportion of repetitive DNA, reassociation will be faster, with higher degree of reassociation

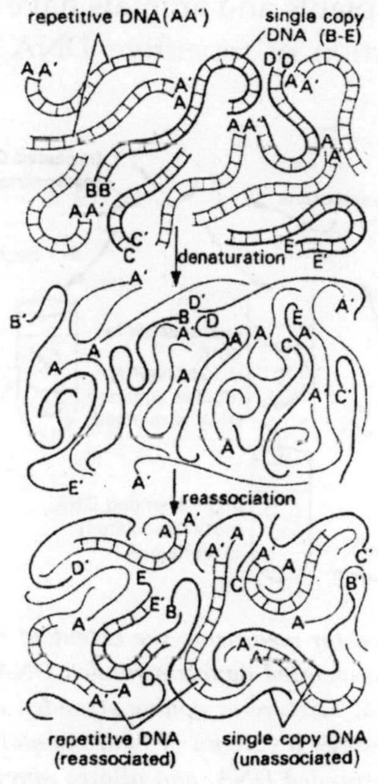

Denaturation and reassociation of DNA containing six copies of repetitive DNA (AA′) shown by thin lines and one copy each of four different unique or single copy DNA's (B-B′; C-C′; D-D′; E-E′) shown by thick lines. Note that more repetitive DNA has reassociated than the single copy DNA. This principle is used for determining the proportion of repetitive DNA in a genome.

achieved at lower *Cot* values. Reassociation rate will also depend on genome size, because in a small genome, higher degree of repetition will be available at the same DNA concentration. Therefore, bacterial DNA with small genome size will reassociate faster, even though no repetitive DNA is found there, so that the range of *Cot* values for complete reassociation is small. In eukaryotes with a bigger genome size, on the other hand, the range of *Cot* values is large suggesting a longer time required for reassociation. But a specific fraction of DNA reassociates faster, indicating the presence of repetitive DNA.

The patterns of reassociation in a number of organisms are shown in *Cot* curves in Figures. During the last more than 25 years, hundreds of plants and animals have been analysed for the relative proportion of repetitive DNA in their genomes (Table).

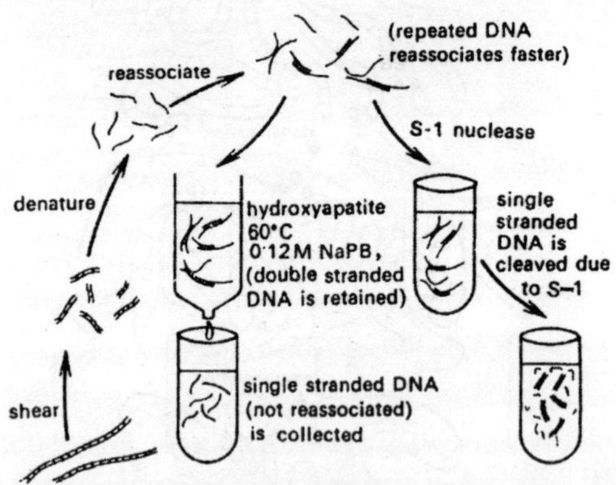

Technique used for measuring the extent of reassociation by separating reassociated double stranded DNA from remaining single stranded DNA. Mixture of double stranded and single stranded DNAs is passed through a column of hydroxyapatite crystals, which retains the double stranded DNA, and allows single stranded DNA to pass through. Double stranded DNA can be eluted by raising temperature and its quantity measured.

Relative Proportions (per cent of haploid genome) of Repetitive Sequences (values are approximate).

Organism			*Proportion of repetitive sequences (%)*
I	Bacteria		—
II	Animals	Sea urchin	38
		Snail	60
		Toad	45
		Mouse	40
		Calf	45
		Human	30

Contd....

Organism		*Proportion of repetitive sequences (%)*
III Plants	Bread wheat	80
	Oats	83
	Maize	78
	Garden pea	75
	Barley	76
	Rye	92

Chemical Complexity vs Sequence (Kinetic) Complexity : Renaturation of DNA in solution as discussed above, depends on random collisions between the complementary strands, which follows second order kinetics, since concentrations of both (not only one of them) the complementary strands will influence the rate of reaction. The rate of reaction, when expressed through differential calculus is given by

$$\frac{DC}{dt} = -KC^2$$

(C = concentration of single stranded DNA at time *t*)
(*K* = reassociation rate constt.)

By integrating the above equation (consult a book on Integral Calculus, if necessary), between the limits of C_0 at t = 0 and C_0 at time $t_{1/2}$ (when reaction is half complete), we get the following expression, ($C_0t_{1/2} = Cot_{1/2}$, which means Cot value at which reassociation is half complete).

$$C/Co = \frac{1}{2} = \frac{1}{1 + KC_0t_{1/2}}$$

or $1 + KC_0t_{1/2} = 2;\ C_0t_{1/2} = \frac{1}{K};\ K = \frac{1}{C_0t_{1/2}}$

In view of the above, reassociation of any particular DNA can be described by the rate constant K (in units, nucleotide moles liter^{-1} sec^{-1}) or in the form of its reciprocal i.e. $C_0t_{1/2}$ (in nucleotides moles × sec/liter). It is obvious that reassociation depends on concentration (Co) and time of incubation (*t*), *i.e.* *Cot* as described above. A higher $Cot_{1/2}$ means slower reaction and lower $Cot_{1/2}$ meant faster reaction. This parameter $Cot_{1/2}$ varies for different organisms and also for different DNA fractions from the same organism.

The $Cot_{1/2}$ of a reaction will be directly proportional to the DNA content, if there is no repetition of sequences within a single genome. For instance, if the bacterial genome has 0.004 pg DNA (4.2×10^6 bp) and if Co is 12 pg per unit volume, it will have 3000 copies of each DNA sequence. Contrary to this if a eukaryotic genome is 4 pg, then there will be only 3 copies in 12 pg *i.e.* concentration is 1000 times less than in bacterial genome, provided there is no repetition of sequences in the genome of 4pg.

In view of this, $Cot_{1/2}$ will indicate the length of all the different sequences (represented only once) in a genome, which will be less than the length of total DNA in a genome when there is repetition. This will be described as kinetic complexity of the genome, and can be determined by knowing $Cot_{1/2}$ and comparing it with $Cot_{1/2}$ of a genome with known complexity (*e.g. E. coli,* which has a genome = .004 pg DNA = 4.2×10^6 bp with $Cot_{1/2} = 9$). The following relationship is used for calculating the kinetic complexity.

$$\frac{\text{Cot}_{1/2}\text{ of DNA of any genome}}{\text{Cot}_{1/2}\text{ of } E.coli \text{ genome } (=9)}$$

$$= \frac{\text{Kinetic complexity of genome}}{4.2\times10^6 \text{ bp}}$$

$$\text{Kinetic complexity of genome} = \frac{4.2\times10^6}{9} \times \text{Cot}_{1/2}\text{ of genome}$$

Figure shows reassociation curves of several DNAs, which are all kinetically pure, meaning that DNA is homogeneous from the point of view of complexity. In eukaryotes, the genome contains more than one such pure components as shown in the comparison made in Figure. In this figure calf DNA has been

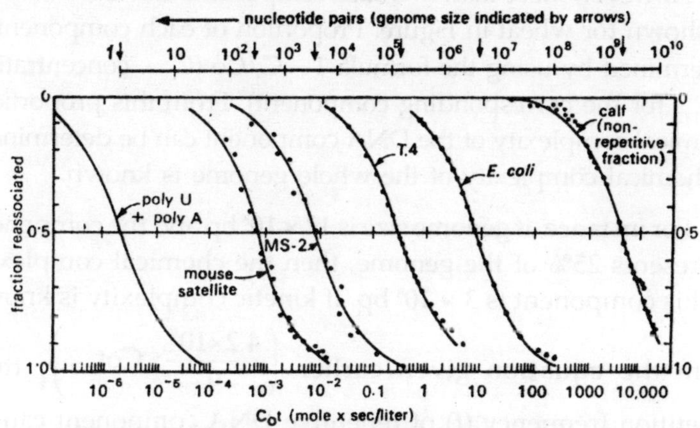

Reassociation patterns of double stranded DNA from various sources. All curves are typical S-shaped indicating homogeneity, but they are displaced to different values of Cot due to genome size.

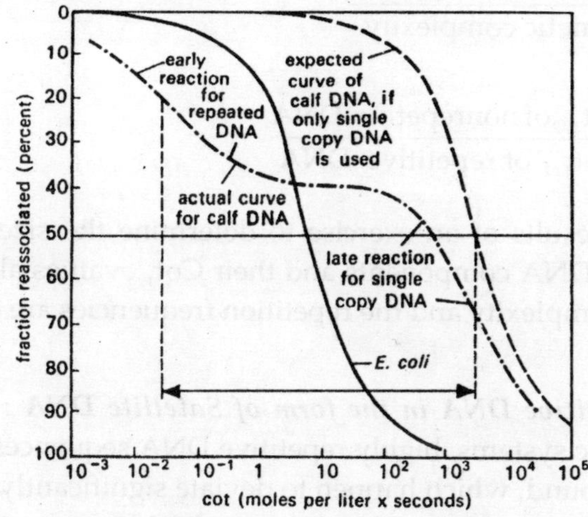

Reassociation patterns of bacterial and calf DNA. Note that the reaction on calf DNA takes place in two stages, one early representing repetitive DNA and the other representing single copy DNA. Midpoints of these two stages are indicated by dotted vertical lines, separated by a factor of 1000,000 (for details see text)

shown to have two components, each with its characteristic $Cot_{1/2}$ value.

However, more than two such components can also be found as shown for wheat in Figure. Proportion of each component is determined by using the formula 1 – C/Co (C = concentration at $t_{1/2}$ for the corresponding component). From this proportion, chemical complexity of the DNA component can be determined, if chemical complexity of the whole genome is known.

For instance, if genome size is 12×10^8 bp and the component represents 25% of the genome, then the chemical complexity of this component is 3×10^8 bp. If kinetic complexity is known from the equation given earlier $\left(\frac{4.2\times10^6}{9}\times Cot_{1/2}\right)$, then repetition frequency (f) of repetitive DNA component can be determined using the following formula :

$$f = \frac{\text{chemical complexity}}{\text{kinetic complexity}}$$

$$= \frac{Cot_{1/2}\text{of nonrepetive DNA}}{Cot_{1/2}\text{ of repetitive DNA}}$$

The results of an exercise to determine the size of three different DNA components and their $Cot_{1/2}$ values along with kinetic complexity and the repetition frequencies are shown in Figure.

Repetitive DNA in the form of Satellite DNA : In many eukaryotic systems, highly repetitive DNA sequences of small size are found, which happen to deviate significantly from the average base composition of the total DNA.

Such a fraction of DNA will have a distinct buoyant density (measured through density centrifugation), which may be higher or lower than that of the total DNA. Therefore, this fraction can be separated as a band distinct from the main band of DNA on $CsCl_2$ density ultracentrifugation and is called the satellite band (Fig.). The satellite band is found on the left of the main band if lighter and on the right if heavier than the DNA of the main band. The DNA separated as a satellite band on ultracentrifugation is therefore, called satellite DNA and has been identified in a variety of eukaryotes including mouse, rat, guinea pig, man, fruitfly, and several dicotyledonous flowering plants. There can also be more than one satellite bands in the same organism, and in such a case they are

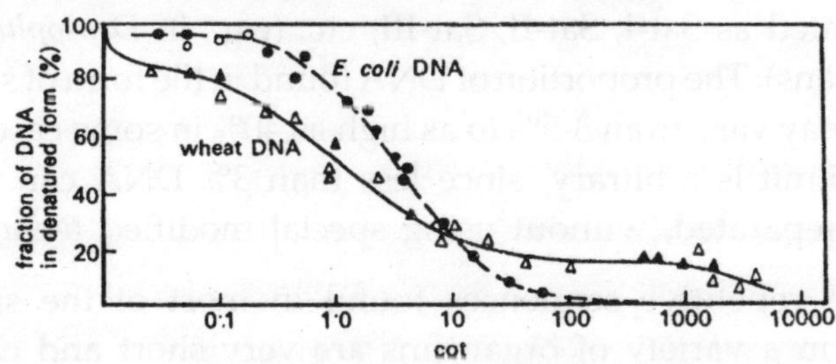

Reassociation pattern of wheat genomic DNA showing more then two homogeneous components (fast, intermediate and slow reassociation).

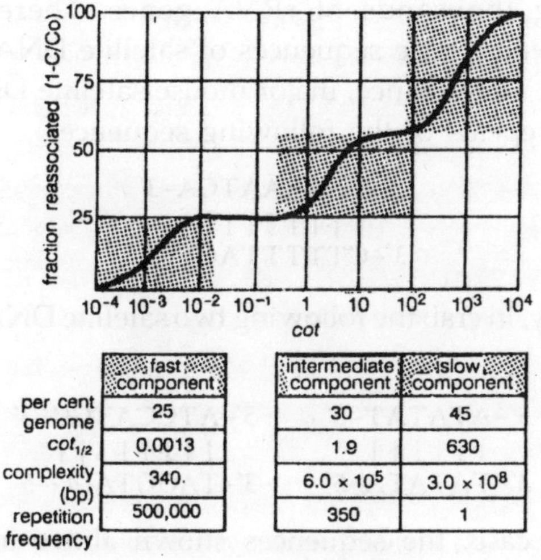

	fast component	intermediate component	slow component
per cent genome	25	30	45
$cot_{1/2}$	0.0013	1.9	630
complexity (bp)	340.	6.0×10^5	3.0×10^8
repetition frequency	500,000	350	1

Reassociation kinetics of eukaryotic DNA, showing calculation of complexity and repetition frequency.

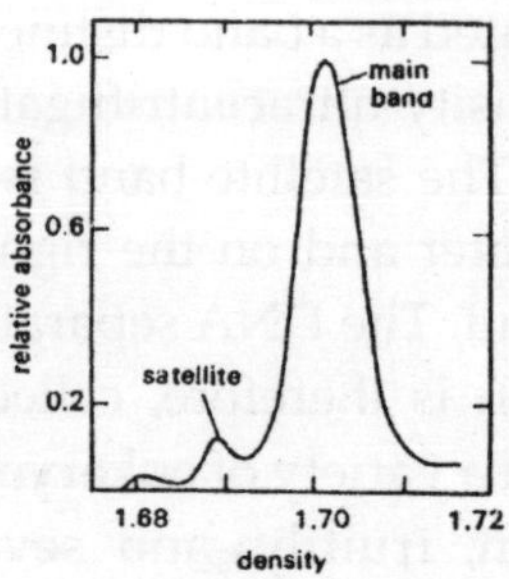

Density centrifugation of embryonic DNA of Drosophila melanogaster, showing a main band and a satellite band.

designated as Sat-I, Sat-II, Sat-III, etc. (e.g., in *Drosophila* and in humans). The proportion of DNA found in the form of satellite DNA may vary from 3-5% to as high as 40% in some cases. The lower limit is arbitrary, since less than 3% DNA can not be easily separated, without using special modified techniques.

The repetitive sequences found in most of the satellite DNAs in a variety of organisms are very short and can not code for any protein. Their function is largely unknown except for ribosomal DNA found in the form of satellite DNA and constituting thousands of rRNA genes. There are many examples, where base sequences of satellite DNA have been worked out. For instance, major mouse satellite DNA consists of tandem repeats of the following sequence.

5′–GAAAAATGA–3′
| | | | | | | | |
3′–CTTTTTACT–5′

Similarly, in crab the following two satellite DNA sequences are found.

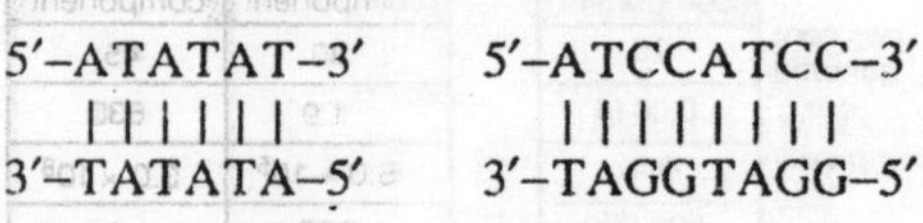

In each case, the sequences shown above are repeated millions of times. In some other cases the sequences in satellite DNA could be as long as few hundred base pairs, as in humans.

In such cases, sometimes, these repetitive satellite DNA sequences may not be very different from some other dispersed repetitive sequences, although in being closely clustered in tandem arrays, they differ from repetitive sequences like *Alu* elements found in mammals. Two remarkable features of satellite DNA are (i) remarkable relative uniformity within the same species and (ii) great variability between otherwise closely related species. In view of these properties, it is difficult to assign any function to satellite DNA except in ribosomal RNA genes as mentioned above.

In-situ Hybridization of Sat DNA : The satellite DNA often lies in heterochromatic regions of chromosomes and its location can be easily demonstrated by the technique of *in-situ* or cytological hybridization. In this technique, DNA within the cell is denatured by treating the cells, that have been squashed on a cover slip. The cover slips can then be incubated in a solution of radioactively labeled satellite or any other DNA, whose location is to be determined. The sat-DNA will hybridize with denatured DNA of the cell and location can be determined by autoradiography. As shown in Figure, position of mouse sat-DNA at the chromosome ends was revealed by *in-situ* hybridization. These chromosome ends also represent the centromeres in mouse.

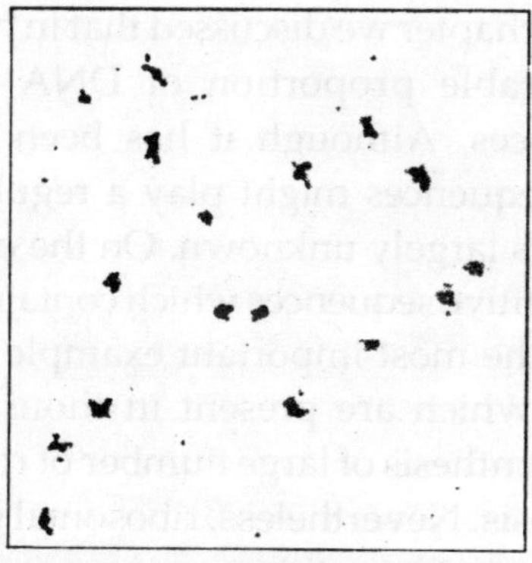

In-situ hybridization of mouse satellite DNA showing its location near centromeres.

Squash Dot Hybridization : The *in-situ* hybridization technique is often used for locating the position of the repeat sequence (including sat-DNA) on a specific chromosome. However, in some cases the repeated DNA from an alien species is used for detecting the presence of chromatin material introduced from this species in a crop plant due to introgression. Sometimes, a cytogeneticist may need to screen large population of a crop species to select plants that may carry alien genetic material. Squash dot hybridization technique is designed for this purpose.

In this technique one root tip from a germinated seed may be squashed onto nitrocellulose filter and the filters are hybridized with radioactively labelled repeated DNA probe derived from the alien species in question. After washing the filters in appropriate buffers, they are used for exposure of X-ray film, which on developing gives the desired information about the presence of alien genetic material. The successful use of this technique has actually been demonstrated to detect the presence of introgressed rye chromatin in several wheat cultivars, through the use of highly repeated DNA probes derived from rye.

Selfish DNA

Earlier in this chapter we discussed that in higher organisms there is considerable proportion of DNA represented by repetitive sequences. Although it has been speculated that these repetitive sequences might play a regulatory role, their definite function is largely unknown. On the other hand, there may be some repetitive sequences which contain definite genetic information and the most important example of such genes is ribosomal genes which are present in thousands within the same cell for the synthesis of large number of ribosomes needed for protein synthesis. Nevertheless, ribosomal genes make such a small proportion of repetitive sequences, that they alone cannot explain the occurrence of these sequences.

In 1980s, it was suggested that these repetitive sequences may not confer any definite advantage to the organisms which contain them. Therefore, sometimes it is said that repetitive sequences represent the DNA which is a repository from which new functional sequences can be obtained. There is, however, no evidence to prove or disprove this hypothesis and it may not be necessary to argue that these DNA sequences, if not of definite advantage to organism, should have been eliminated through natural selection.

Actually, the cells themselves provide an environment and the repetitive sequences may be subjected to selection pressure within the cell. It can, therefore, be argued that these sequences may be maintained through non-phenotypic selection, and by a mechanism directed for the self preservation of these sequences. It is in this context that the term selfish DNA, was used for this repetitive DNA.

One of the examples of selfish genes is found in transposable genetic elements in prokaryotes. These transposable genetic elements are not indispensable and can be transposed from one place to the other or from one organism to the other.

It is difficult to find reason for selection of these sequences represented by transposable genetic elements on the basis of phenotypic fitness. On the other hand, it has been argued that the transposable genetic elements are maintained due to their property of transposability which permits them to escape the pressure of selection exercised on the organism.

Such transposable elements are also found in case of eukaryotes, for example Tyl element in yeast and the copia like and inverted repetitive elements of *Drosophila*. We have already described selfish genes like *SD* in , 't' in mouse, and *M* in flour beetle.

The term selfish genes was used there for those genes, which are successful in propagating themselves, while being detrimental to the organisms that carry them. Therefore, a distinction needs to be made between these selfish genes described earlier and selfish DNA described in this chapter; here it refers to repeated sequences with unknown function.

Packaging of DNA

In any living system, the length of the DNA always greatly exceeds the dimensions of the compartment (e.g. nucleus), which contains it. The DNA (or RNA, as in some viruses) must, therefore, be compressed very tightly to fit into the space available. The density of DNA is about 10 mg/ml in bacteria, about 100 mg/ml in eukaryotic nucleus, and more than 500 mg/ml in T4 head.

The overall compression of the DNA has also been described as packaging ratio (the length of DNA divided by the length of the unit that contains it). This packaging ratio in eukaryotes reaches 1000-2000 in interphase nuclei and up to 10,000 in metaphase nuclei.

The patterns of this packaging and changes in these patterns during replication and transcription (gene expression) have been the subject of detailed study during the last more than three decades. These patterns of packaging are facilitated by associated proteins, which in eukaryotes, could be histones or nonhistones (RNA polymerase is also a nonhistone). More stable proteins in chromosomes are histones and are classified into five types called H1, H2A, H2B, H3 and H4.

Another histone protein H5 (a variant of H1) has also been often described, although it may not be involved in packaging. It was thus known that chromatin consists of DNA, five different but stable proteins and a variety of nonhistone proteins.

It will be seen that the progress in the study of chromatin

structure in early years, did not make much headway due to limitation of techniques. Whole mounts of chromosomes, whether within the cell or isolated, could be examined under light or electron microscope. This gave information about gross morphology of chromosomes including single stranded or multi-stranded nature or about the thickness of the ultimate single strands or fibres.

When ultrathin sections of chromosomes were examined under electron microscope, it gave such an incomplete picture of chromatin structure that no satisfactory models could be prepared. Hundreds of speculatory models based on inadequate available data were presented and rejected before they could become known to students of cytology or genetics.

However, in 1970's due to the use of improved techniques, nucleosome model for chromatin structure became available, which has been confirmed over and again during the last more than two decades. Some of the details regarding the techniques used for discovery of the nucleosome subunit, the structural details of this subunit and the higher levels of organization of these nucleosome subunits will be briefly discussed in this chapter.

Techniques Leading to Nucleosome Discovery : Adequate information for constructing suitable models for chromatin structure started coming after 1960 mainly through development of newer techniques. Techniques from biophysics, genetics and electron microscopy were developed in this connection. These techniques will be discussed in brief outlining the kind of information each produced.

X-ray Diffraction and Electron Microscopy : While studying X-ray diffraction patterns of chromatin, M.F. Wilkins and V. Luzzati in the year 1960 observed patterns which suggested repeating units in chromatin. Similar repeats were inferred from X-ray diffraction of a mixture of DNA and histones also. Electron microscopy was also used for a study of chromatin structure. In 1974, A.L. Olins and D.E. Olins reported results of electron microscopy of chromatin obtained from interphase nuclei lysed in water.

Under electron microscope, chromatin fibre appeared as arrays of spherical particles, about 10 nm (1 nm = 10^{-9} m = 10 Å) in diameter, connected by filaments, which were about 2 nm in diameter. These particles were called v (nu) bodies. Later in 1975, P.M. Outdet and others used the term nucleosome for these particles. Electron microscopy was also done on chromatin subjected to nuclease digestion, and the nucleosome structure was confirmed.

The structure of DNA suggests, that DNA alone can not form repeats at intervals of 10 nm, which meant that the association of histones must have conferred the repeating organization to chromatin.

Association of histones is also suggested *firstly*, from the knowledge that histones are positively charged and will associate with negatively charged groups in DNA, and *secondly* from the knowledge that histones perform their function in chromatin by binding to DNA. Therefore, it suggested that repeating units in chromatin result due to association of DNA with histones and that these repeats occur at intervals of 10 nm (1 Å = 10^{-7} mm = 10^{-10} m).

Chromatin Reconstitution Experiments : X-ray diffraction patterns of native chromatin could be compared with those of chromatin, reconstituted from DNA and histones to find out the degree of successful reconstitution.

It was shown by T. Kornberg (T. Kornberg is different from A. Kornberg of DNA replication fame), that 90% reconstitution was achieved, when DNA was mixed with unfractionated histone preparation. However, when purified histones were used, reconstitution failed suggesting that some denaturation of histones takes place during purification, which destroys the compactly folded native conformation of chromatin.

Nuclease Digestion and Regular Distribution of Histones : Nucleases are a class of enzymes which will cause fragmentation of DNA at those sites which are not protected by proteins like histones. Therefore, it was expected that through a study of DNA fragments generated by nuclease treatment, location of the sites of naked DNA in chromatin will be possible and information will thus be available about the nature of association of histones with DNA in the chromatin. This technique of nuclease digestion was utilized in 1973 by Hewish and Burgoyne for fragmentation of chromatin of rat liver cells.

The fragments were separated by gel electrophoresis. It was found that fragments of different sizes were produced which formed distinct bands representing sizes which were multiples (two, three, four times, etc.) of size of the smallest unit. This information came from further detailed study of bands (Figs.).

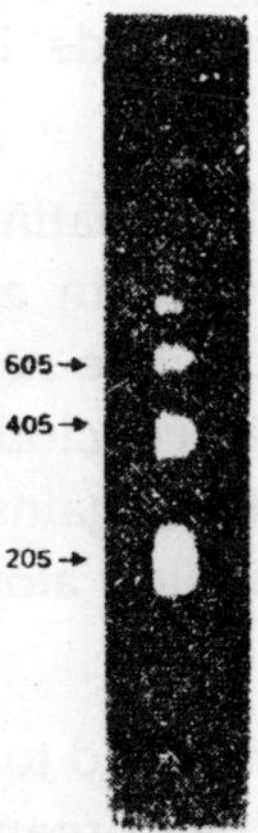

Multimeric series of DNA bands after nuclease digestion of chromatin followed by gel electrophoresis.

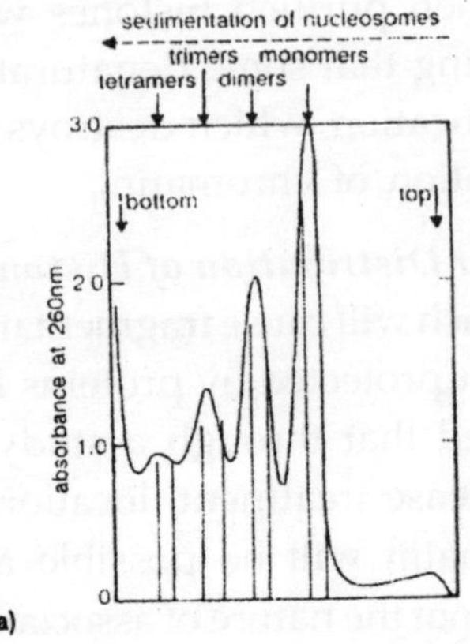

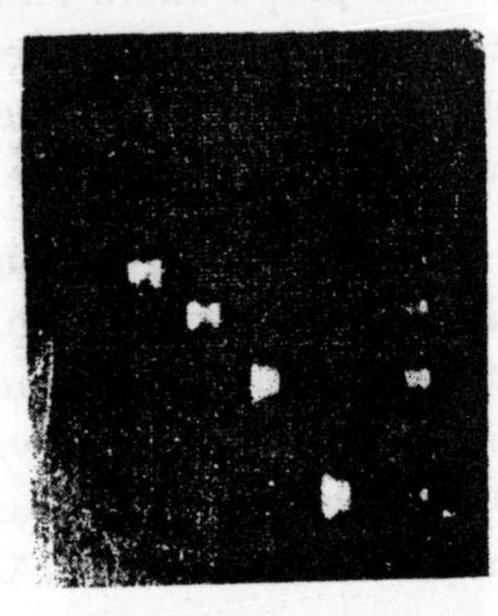

(a) Results or sedimentation of nucleosome preparation on sucrose gradient showing separation of monomers, dimers, trimers and tetramers. (b) Gel electrophoresis of DNA purified from fractions in (a) above suggesting that each multimer had a multiple of unit DNA length, (redrawn from Lewin's 'Genes VI').

On the other hand, when DNA alone and not the chromatin, was fragmented with same nuclease enzyme, uniform smear rather than bands were observed on gel. Appearance of uniform smear rather than bands meant that fragments obtained due to nuclease digestion of naked DNA, gave a gradient of size, rather than being multiples of the size of smallest unit as shown through distinct bands in nuclease digestion of chromatin.

These studies of rat liver chromatin clearly demonstrated that rat nuclease cleaved chromatin at sites spaced at regular intervals along the DNA, while naked DNA was cleaved at random. Therefore, it was concluded that histones confer a regular pattern of protection against nuclease and, therefore, proteins should be distributed along the DNA in a regular periodicity.

Since, histones were believed to be distributed along DNA in a regular periodicity, their organization in chromatin, was an important aspect for study. Some information about this organization came with the help of gel filtration technique.

This technique is just opposite to gel electrophoresis, because in gel electrophoresis, heavier molecules move slowly, while in gel filtration, heavier molecules move faster.

In gel filtration, small molecules can enter gel particles and their movement, therefore, is obstructed, but the bigger molecules pass around the particles and thus move more rapidly down the column. When histones were subjected to gel filtration, these were clearly separated into two groups, one consisting of H1, H3 and H4 and the other group consisting of H2A and H2B.

This was considered to be surprising, because H1 was known to be double the size of the other proteins, and therefore, should have moved faster and separated itself from others. Therefore, it was speculated that H3 and H4 could associate into a dimer to have a size equal to that of H1.

Distribution of Histones within a Repeat unit Through Cross Linking : There are certain chemicals which help in chemical cross linking and thus facilitate in the study of cross linking between molecules, if any. When this technique was applied to histones in solution, it was discovered that H3 and H4 were present as a tetramer (or a pair of dimers) i.e. $(H_3)_2.(H4)_2$ and not as a simple dimer (H3.H4).

This tetramer nature was subsequently confirmed in chromatin also. The relationship of this tetramer with other histones and DNA was also studied. It was shown that chromosomes of most organisms had equal number of molecules of histones H2A, H2B, H3 and H4. It was also shown that 25 nucleotides of DNA were present, per histone molecule.

Since H3 and H4 are present as tetramers, $(H3)_2.(H4)_2$, each repeating unit may have one such tetramer (described as kernel) and two molecules each of H2A and H2B (because in chromatin H2A = H2B = H3 = H4). Histones will thus form an octamer,

Properties of a Nucleosome Model

Different constituents	*Nuclcosome*		*Core particle*	
	Size	*Mass (daltons)*	*Size*	*Mass (daltons)*
1. DNA	200 base pairs (67 nm)	1,30,000	140 base pairs (67nm)	90,000
2. Histories (H2A, H2B, H3, H4)	Octamer (2 molecules of each)	1,08,000	Octamer (2 molecules of each)	1,09,000
3. HI Histone	1 molecule	24,000	Absent	
	Total mass	2,62,000		1,99,000

$(H3)_2.(H4)_2/(H2A.H2B)_2$ associated with 200 bases in each repeating unit. One molecule of H1 is also associated with each octamer in a nucleosome unit. However, H1 was not linked with other histones, so that in core particle H1 was absent.

Subunit of Chromatin—The Nucleosome

Basic Structure of a Nucleosome : On the basis of X-ray diffraction, electron microscopy, nuclease digestion, and chemical cross linking, a model for necleosome structure was proposed. Properties of this nucleosome model are presented in Table and Figure. The shape of nucleosome, arrangement of histones including H1 were also studied.

The diameter of nucleosome is 11 nm and its height is 6 nm. The length of DNA arourd it is 70 nm (~ 67 nm DNA = approx. 200 base pairs, because 34 Å = 3.4 nm = 10 base pairs), which is equivalent to 200 base pairs. The length of DNA varies not only in different organisms (from 154 bp in a fungus to as high as 260 bp in a sea urchin sperm), but also in different tissues or even in different regions within the same cell of an organism.

The length of DNA, however, suggested that DNA should be coiled around nucleosome core particle. Enzyme DNAase I also led to cleavage of DNA suggesting that DNA lies on the surface of nucleosome. It was, therefore, suggested that histones do not cover DNA, but DNA is wrapped around histone core particle.

In other words, it was suggested that it will be more appropriate to describe chromatin as string on beads rather than as beads on string. Neutron scattering technique was also used to find out distances of histones and DNA from centre of nucleosome. It was shown that DNA was farther away than proteins from the centre, again suggesting that

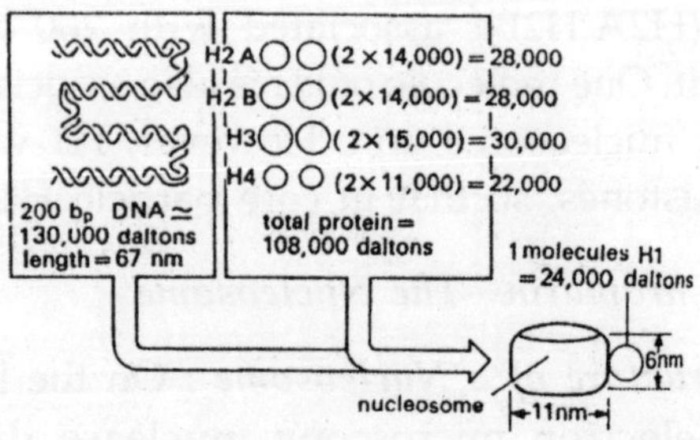

Different constituents (DNA and histone proteins) of a nucleosome and their properties.

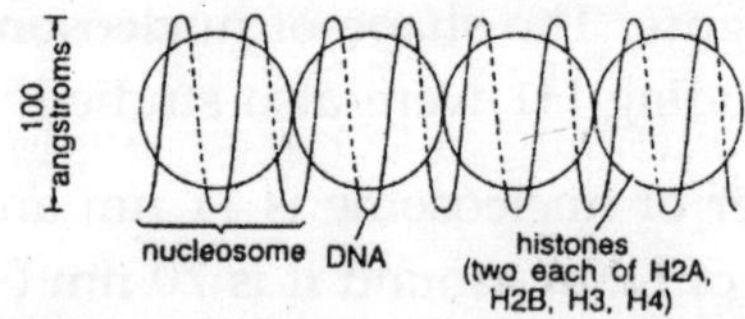

Original proposal for nucleosome repeating units, with DNA wound on a series of beads. Actual shape of histone complex or the path of DNA was not known initially (redrawn from Sci. Amer., Vol. 244, 1981).

DNA is not covered by histones, but is wrapped around histone proteins. A simple model of chromatin showing DNA wrapped around histone core is shown in Figure.

Structural Periodicity and Super-coiling of DNA in a Nucleosome : The structural periodicity of DNA coiled around the histone octamer, was examined by treatment with enzymes

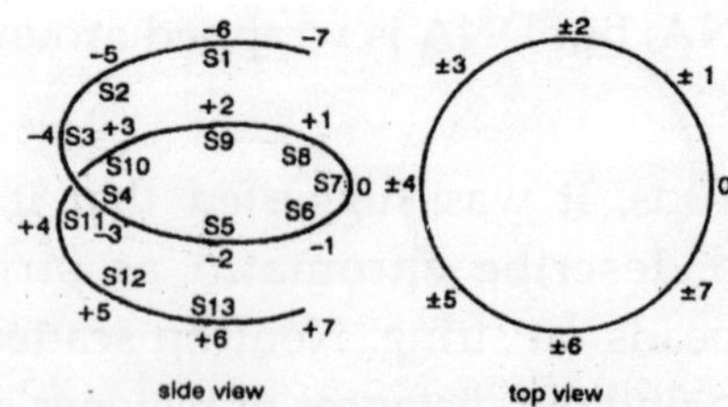

Two numbering schemes dividing core particle DNA into segments, 10 bp apart; SI to S13 are cleavage sites for DNAase I and DNAase II.

DNAase I and DNAase II. Both these enzymes make single-strand nicks in DNA, so that double stranded DNA structure is not disturbed.

But on denaturation following nicking, single stranded DNA fragments are released. It was shown that these cleavage sites are ~ 10 bases apart and are designated as S1 to S13 (S1 is ~ 10 bases away from 5' end, S2 is ~ 20 bases away from 5' end, and so on; see Fig.).

Not all sites are cut with equal frequency, so that on electrophoresis a ladder with segments of lengths, which were multiple of 10 were obtained. It was also shown that pairs of sites from S1 to S4 or S10 to S13 lie 10 bp apart, while those from S4 to S10 lie 10.7 bp apart.

The above results differ from cleavage of DNA immobilized on a flat surface, where the cutting periodicity reflects structural periodicity (10.5 bp per turn). A variation in cutting periodicity in core DNA of nucleosome (10.0 bp at ends; 10.7 bp in the middle) means that there is variation in structural periodicity of core DNA; in the middle, this is close to the value of DNA in solution, but at the ends it is less tightly screwed.

Much work on the nucleosome has also been conducted in virus SV40 having a circular molecule of DNA, 5200 bp long (1500 nm long). The DNA is packed into a series of nucleosomes, collectively called a mini-chromosome, which is 210 nm long showing a packing ratio of about 7.

This mini-chromosome has been used for a study of supercoiling of DNA in nucleosomes. The DNA in chromatin undergoes supercoiling due to the following reasons: (i) DNA is wrapped around the nucleosome octamer. (ii) Nucleosomes are also coiled using 5-6 nucleosomes per turn. Some of this supercoiling is restrained by the associated proteins (histone octamer). Theoretically, the path of DNA around one

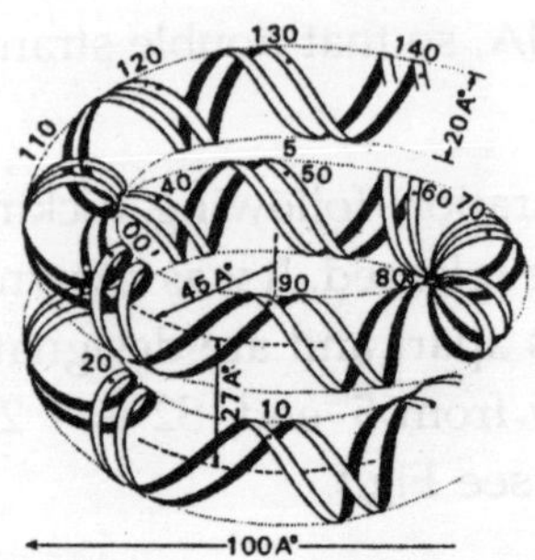

Path of DNA superhelix on a nucleosome core. There are about 80 nucleotide pairs of DNA per turn, one and three quarter turns of DNA are wrapped on histone complex (redrawn from Sci. Amer., Vol. 244, 1981).

nucleosome should generate –1.8 supercoils, if no strain is exercised by proteins. Using SV40 DNA, measurements of supercoiling in DNA involved in nucleosome structure were made which gave a value of –1 turn per nucleosome as against –1.8 expected in the absence of proteins. This discrepancy is sometimes described as linking number paradox (linking number = number of times the two DNA strands are intertwined in circular DNA). The periodicity of 10 bp in DNA wrapped on nucleosome is actually lower than 10.5 bp found in DNA' in solution. This allows for absorption of a part of negative supercoils reducing it from –1.8 to –1.0 in the nucleosomes.

Core Particle : When nuclease treatment was prolonged beyond the cleavage between nucleosomes, the DNA was removed from both the free ends, yielding particles containing DNA of only 146 (or sometimes 140) nucleotide pairs instead of 200 nucleotide pairs (Fig.). This reduced form of nucleosome is called core particle. Due to nuclease treatment, the length of the DNA is reduced in discrete steps as shown in Figure. Adjacent core particles, each with core DNA, are joined with each other through DNA segment called linker DNA, which is removed during prolonged digestion. The core DNA has an invariant length of 146 bp, but the linker DNA can vary from 8 bp to 114 bp, thus accounting for variation in DNA length of nucleosomes (154-260 bp).

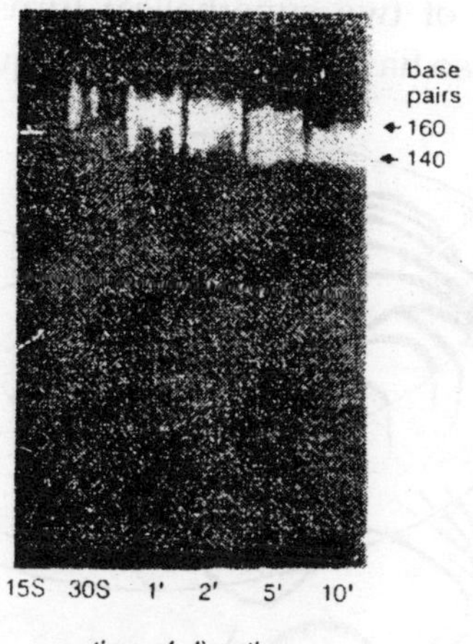

Reduction in the length of nucleosome DNA by nuclease digestion in time (15 seconds to 10 minutes)

Low Resolution (7Å) Structure of Core Particle : In the year 1977, the core particle was available in crystalline form for a study of X-ray crystallography (conducted only in 1984) to reveal the three dimensional structure. Electron microscopy and X-ray diffraction patterns also revealed that nucleosome core particle is neither spherical, nor perfectly flat, but is wedge shaped, about 11 nm in diameter and 6 nm high. The study of crystals suggested that DNA double helix is coiled into a large helix or superhelix, by making two turns around the histones in the middle of the particle. The two turns of the DNA double helix will almost touch, since DNA is 2 nm thick and the particle is only 6 nm thick. From the structure of DNA and the diameter of core particle, it has been estimated that two superhelical turns would have about 160 base pairs (80 bp per turn). Since length of DNA in a core particle is only 146 pairs, it will occupy only 1.7 turns. The particle, therefore, will be thinner on one side, accounting for the wedge shape. It was also shown that histone octamer itself is also bipartite and wedge-shaped providing for a short helical ramp on which the DNA is wound (Fig.).

The closeness of two superhelical turns of DNA on the histone octamer also has a functional consequence. This allows

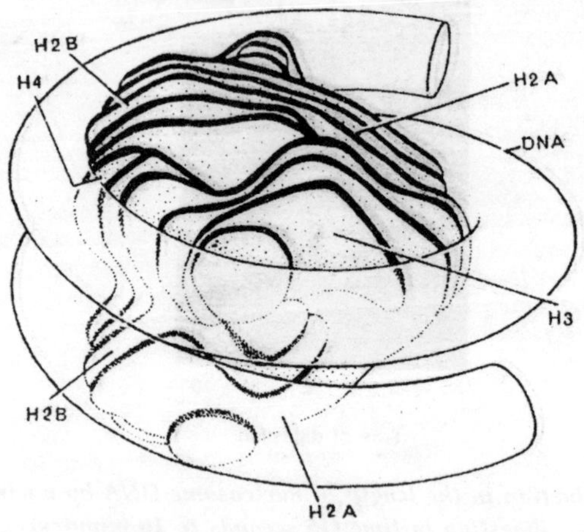

Model of nucleosome core with DNA superhelix wound on histone octamer, which has, a more or less, continuous helical ramp, on which DNA (146 base pairs) is wound.

the binding of a protein molecule simultaneously to two different sites of DNA, 80 bp apart. It will be seen later that a promoter site (CAAT box) may lie 80 bp upstream from start point. In such a situation superhelical turns in the nucleosome may allow the association of RNA polymerase enzyme to the promoter and initiation sites simultaneously due to closeness of two turns on the nucleosome octamer. In 1993, in a study of interaction of nucleosome with *Xenopus* vitellogenin Bl promoter, it was shown that the presence of nucleosome stimulates transcription by oestrogen. It is believed, that by DNA looping, the oestrogen receptor binding site is brought closer to the site of transcription complex formation.

High resolution (2.8Å) structure of core particle. In the year 1997, higher resolution *(2.8Å)* of nucleosome structure was obtained, showing how the nucleosome core is assembled (Fig.). It was shown that each histone consists of (i) a structured three-helix domain called histone fold and (ii) two unstructured tails. The histone fold helps to form heterodimers (H2A-H2B; H3-H4) each through a hand-shake like arrangement and these dimers together form the histone octamer. This histone fold has also been observed in a number of transcriptional coactivators and repressors, providing an evolutionary link

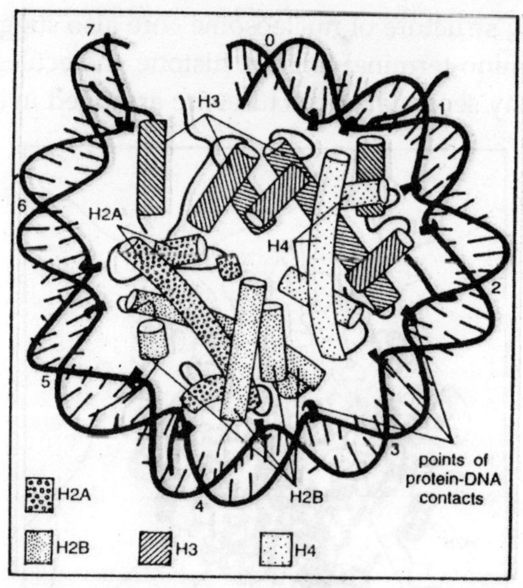

Arrangement of histones and DNA in a core particle of a nucleosome as revealed at 2.8A° resolution.

between chromosome structure and the transcriptional machinery. It was shown that 121 bp of the total 146bp of DNA are organized by the histone-fold domains. Following are some of the characteristic features of this organization: (i) At the sites of DNA's entry into and exit from the nucleosome, the DNA is bound by extensions of the H3 histone fold; (ii) each heterodimer binds about 30bp of DNA, with contacts at l0bp intervals; (iii) the phosphate backbone of the DNA makes many contacts with the atoms of the main chains of the histone proteins; (iv) arginine side chains peretrate all the 14 minor grooves (in a sequence of 146bp, there will be 14 minor grooves) that face the histone proteins; (v) the path of DNA superhelix is not uniform, but distorted through bends (generated by H3-H4 and H2A-H2B) at several positions, and the average number of base pairs per turn ranges from 10.6bp (in solution) to 10.2bp; this allows the neighbouring turns of DNA superhelix to line up forming channels for histone tails.

The new structure of nucleosome core also suggested that the long amino-terminal-tails of histone molecules, without assuming any secondary structure, are arranged in an orderly

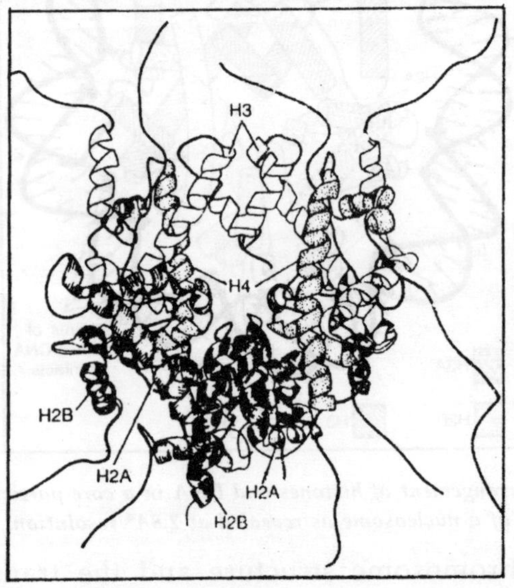

The arrangement of long amino terminal tails of different histone molecules of a core particle of a nucleosome as revealed at 2.8A° resolution.

manner and are involved in the following nucleosome-nucleosome interactions (Fig.): (i) tails from H3 and H2B pass through minor-groove channels, thus protruding out of DNA surface every 20bp; tails of H3s- extend beyond DNA surface, coming in contact with another nucleosome; (ii) one of the two protruding tails of H4 makes many contacts with H2A-H2B dimer of a neighbouring nucleosome; (iii) acetylation of histone tails is associated with transcriptional activity and underacetylation is linked with repression.

Spatial Arrangement of Histones

Arrangement of four histones in the nucleosome was also studied through a study of specific sites in DNA and of their interaction with four histones. These sites could be assigned to specific regions of octamer and, therefore arrangement of eight histone molecules could be described. It was shown that $(H3)_2.(H4)_2$ tetramer defines the central turn of the DNA superhelix. H2A and H2B are added as two (H2A).(H2B) dimers one on each face of the tetramer $(H3)_2.(H4)_2$ and each binding additional DNA and completing the superhelix (Fig.). $(H_3)_2$

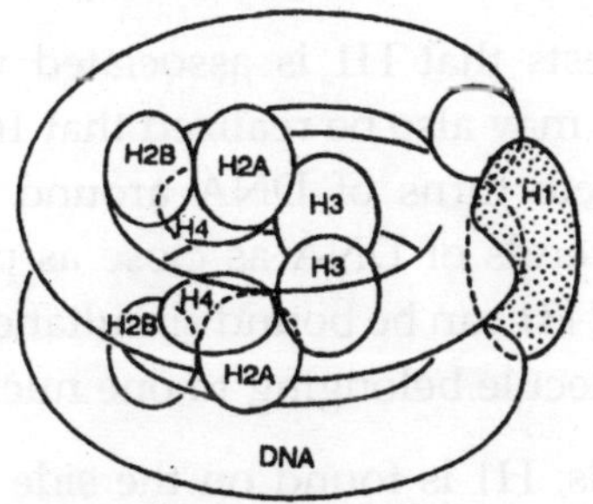

A model for nucleosome with $(H3)_2.(H4)2$ forming a kernel.

$(H_4)_2$ tetramer makes a central 'kernel' associated with two independent dimers H2A-H2B. This explains, why H3 and H4 alone can confer nucleosome like properties on DNA but H2A and H2B can not confer this property.

Relation Between Different Nucleosomes : The relationship between different nucleosomes was examined on the basis of results of digestion of chromatin by micrococcal nuclease. We have already discussed that the first effect of nuclease is cleavage between nucleosomes which, on prolonged digestion, are reduced to core particles. At an intermediate stage before the core particles are produced, DNA was found to be 166 base pairs long and not 146 base pairs long as in the core particle. When this intermediate stage with 166 base pairs is further degraded to core particle, the histone protein H1 is assumed to fall off due to removal of 10 base pairs from each end of

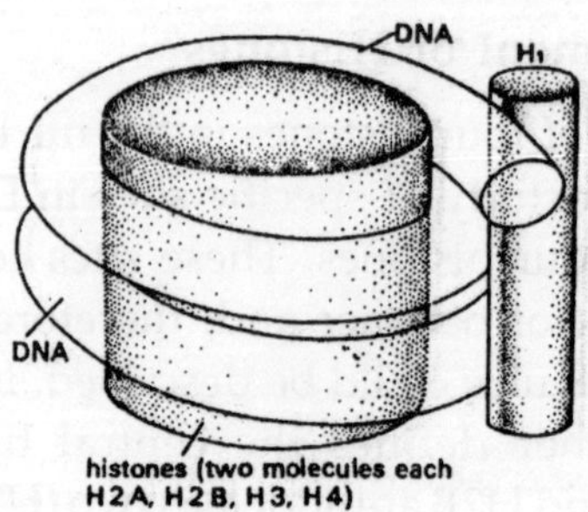

Model of full nucleosome (not the core particle) showing the role of H1 histone, which is attached at the entry and exit of two turns of 166 base pairs long DNA

DNA. This suggests that H1 is associated with the ends of DNA molecule. It may also be realized that 166 base pairs will make two complete turns of DNA around the nucleosome, bringing the two ends of DNA as close as possible so that a single molecule of H1 can be bound simultaneously to both the ends of DNA molecule belonging to one nucleosome subunit.

In other words, H1 is found on the side of a nucleosome at regions of entry and exit of DNA superhelix, as shown in Figure. Since H1 is not necessary for coiling of DNA, it is only an accessory protein and may be involved in condensation of chromatin.

Solenoid Model

It was also shown that 11 nm wide fibre of nucleosomes gets coiled upon itself to form ~ 30 nm wide helix with five or six nucleosomes per helix. In this helix successive nucleosome units came close together, so that their centre to centre distance was about 10 nm. This 30 nm structure was called a solenoid (term solenoid in physics is used for a wire coiled on a central axis). Formation of solenoid from nucleosomes can be compared with winding of a cable (DNA) on a spool (nucleosome) and then folding of wrapped spools. A suggested structure of solenoid is shown in Figure.

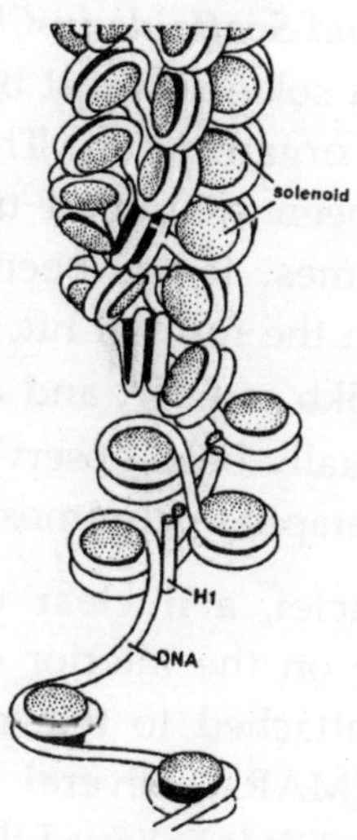

Solenoid model, with a nelix having about six nucleosomes per turn, where H1 molecules of adjacent nucleosomes are in contact.

It was also proved that H1 protein helped in folding of 110 Å wide fibre into 300 Å wide solenoid, because when H1 was removed, this ordered folding was found to be absent and only irregular clumping of nucleosomes could be observed. It has been shown that H1 molecules aggregate by cross linking to form polymers and may thus control the formation of solenoid. It is also speculated that solenoid has to fold or coil again during condensation of chromatin.

The above account gives patterns of coiling and packing of DNA. Since 60 nm (600 Å) long DNA (166 base pairs) is coiled in a nucleosome, only 6 nm (60 Å) long, and then nucleosomes are coiled in 30 nm (300 Å) wide solenoid fibres, it gives DNA a packing ratio of 1 : 50. However, in highly condensed chromosomes, the packing ratio is actually 1 : 5,000, which is 100 times greater than provided by solenoid model. This 100 times packing of the solenoid, should take place by further coiling and folding of solenoid.

Loops, Domains and Scaffolds in Chromatin : The 30 nm fibres in the form of a solenoid must be folded in structures at a higher level of organization. These higher levels of organizations have been examined using metaphase and interphase chromosomes. It has been shown by DNAase digestion that DNA in the form of nucleosomes is organized in loops, each with 85kb of DNA and a length of 10-30 μm. These loops have actually been observed to emanate from a central scaffold in metaphase chromosomes.

In interphase nuclei, a nuclear matrix is found as a filamentous structure on the interior of nuclear membrane. Chromatin remains attached to this matrix through matrix attachment regions (MARs). Several MAR sequences have actually been identified. It is believed that the same sequences of DNA work as MAR in interphase nuclei and as SAR (scaffold attachment regions) in metaphase chromosomes. Both matrix and scaffold are proteinaceous in nature.

Chromatin Replication and Nucleosome Assembly : Having known the organization of DNA in the form of nucleosomes, one may like to ask the question-— what happens to the nucleosomes during replication and how are new nucleosomes assembled ? Cross linking experiments suggested that the old histone octamers are conserved during replication, so that when chromatin is replicated, the existing histone octamers are displaced from DNA to allow its replication. It has also been shown that the conserved octamers may reassociate with either of the two daughter duplexes. An equal number of histone octamers must be synthesized quickly, so that the nucleosomes reform more or less immediately on both the daughter DNA molecules. Although the histone octamer may be conserved, but all the conserved octamers may not reassociate with the same daughter DNA molecule. Instead they may follow a dispersed mode of distribution, in which old and new octamers are intermingled at random during association with the two daughter DNA molecules (Fig.). There is also some support for semi-conservative assembly of histone proteins. Although $(H3)_2.(H4)_2$ tetramers are stable, $(H2A.H2B)_2$ tetramers could result from heterodimers (H2A conserved and H2B newly synthesized or *vice versa).* These two models are not mutually exclusive and both may operate in a single cell.

It has been shown that in the assembly of nucleosome, an assembly protein called nucleoplasmin (a pentamer containing a sub-unit with mass of 29,000 daltons and found in the nucleoplasm) plays an important role. Nucleoplasmin binds to histones (but not to DNA or nucleosomes) and may thus release histone octamer from the DNA in a more regulated way, so that these may not make random aggregates. It is also suggested that the nucleosomes assemble first and the DNA supercoiling follows this assembly.

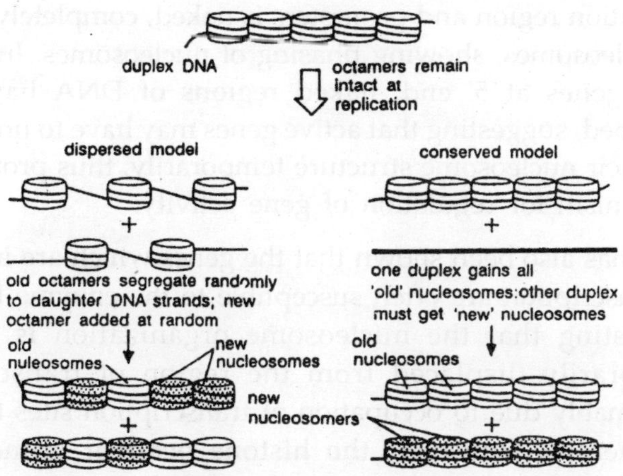

Dispersed and conserved models of nucleosome replication and assembly (dispersed model has been found to hold good).

Phasing and Modification of Nucleosomes in Active Genes : Active and inactive chromatin have been compared with respect to nucleosome organization. It is shown that nucleosome in the form of repeats are present in both cases, but in active form, nucleosomes are displaced away from the regions of DNA to be replicated or transcribed. For this purpose, the histones undergo a variety of modifications and some new non-histone proteins become associated with the chromatin.

Nucleosome Phasing : Some specificity in the distribution of DNA sequences on the nucleosome and a lack of their random distribution is called nucleosome phasing. Recent evidence suggested that nucleosomes are really phased along the length of DNA and that through their position on DNA, regulatory proteins may expose or protect certain DNA sequences. The spacing of nucleosomes is not random in statistical sense, but is rather regular and non-random.

It has also been shown that spacing of nucleosomes may vary in different tissues, suggesting that the locations of nucleosomes are not constant but may vary. In the chromosomes of SV40, a DNA segment of 400 base pairs encompassing the replication region and promoters is naked, completely devoid of nucleosomes, showing phasing of nucleosomes. In several other genes at 5' end, naked regions of DNA have been described, suggesting that active genes may have to uncoil and lose their nucleosome structure temporarily, thus providing a mechanism for regulation of gene activity.

It has also been shown that the genes which are in a state of transcription are often susceptible to the enzyme DNAase suggesting that the nucleosome organization is at least temporarily displaced from the region of transcription, presumably due to occupation of transcription sites by RNA polymerase. However, the histone octamer immediately recaptures its position as soon as the transcription is over.

The susceptibility of chromatin to DNAase is attributed to lose binding of some nonhistone proteins like HMG14 and HMG17 (HMG = high-mobility group), because, once these proteins are released, the chromatin does not exhibit susceptibility. It has also been shown that some hypersensitive sites (digested with very low concentration of DNAase) are found upstream to the initiation site for transcription. The sites are free of nucleosomes and can be 100-200 bp long.

Ubiquitination, Acetylation, Methylation and Phosphorylation of Histones in the Nucleosomes : The histone proteins, which are integral parts of nucleosome undergo a variety of modifications to bring about decondensation of chromatin, to allow access of DNA replication or transcription machinery to naked DNA. These modifications include ubiquitination, acetylation, methylation and phosphorylation of some specific amino acid residues of histones. For instance, the protein ubiquitin, which is ubiquitous for its presence in all living cells, gets associated with H2A in the form of U-H2A in 10-30% nucleosomes. Usually only one of the two H2A molecules of a nucleosome is ubiquitinated. There is a tendency for U-H2A to be concentrated in the nucleosomes of the transcribed sequences.

All histone proteins present in the nucleosomes also undergo modification due to acetylation, methylation and phosphorylation. Acetylation and methylation occur on the free amino groups of lysine residues. Methylation also occurs on arginine and histidine. Similarly, phosphorylation occurs on the hydroxyl group of serine and histidine. Methylation and acetylation remove the positive charge on NH_3^+, while phosphorylation introdues a negative charge in the form of phosphate group. These modifications in histones are transient and are associated with structural changes that occur in chromatin at the time of replication and transcription.

During S-phase (DNA replication and histone synthesis) of cell cycle, histones appear to be activated and methylated, but the acetyl and methyl groups are removed during other parts of the cell cycle. It appears, therefore, that acetylation and methylation help in the nucleosome assembly during S-phase of the cell cycle. Acetylation has also been shown to be involved in the regulation of transcription. For instance, acetylated lysine residues in the NH_2-terminal domains of nucleosomal histones serve as landmarks for transcriptionally active chromatin within the chromosome. Hyperacetylation, also increases the accessibility of nucleosomal DNA to transcription factors. In contrast to this, deacetylation is correlated with transcriptional silencing. Histone acetyltransferases and deacetylases thus play an important role in regulating transcription. Certain deacetylases have been shown to have homology with some transcriptional regulators (e.g. Rpd3p in yeast).

Phosphorylation and dephosphorylation of H1 histone have also been observed during the cell cycle and are known to play a role in chromosome condensation during mitosis. Phosphate groups are added during mitosis and are removed soon after the division is over.

The above modifications in histone proteins also facilitate binding of certain transcription *x* factors many of which are known to be non-histone proteins. For instance, in *Xenopus,* it was shown that placement of a nucleosome (histone octamer) on the 5S gene, blocks binding of the transcription factor TFIIIA. It was also shown that the amino terminal regions of histone molecules are either cleaved off or get heavily acetylated, which allows free access of TFIIIA to nucleosomal DNA. It is also believed that the acetylation either releases the amino terminal tails from DNA, thus making it accessible or it induces a conformational change in the nucleosome, which promotes TFIIIA binding.

As shown above, considerable information is now available about the factors involved in chromatin decondensation that is needed both for DNA replication and transcription.

4

Mitochondrial and Chloroplast Genomes

The nuclear DNA encodes most of the proteins, including those for the organelles. It was shown early that most of the proteins found in mitochondria and chloroplasts are encoded by nuclear DNA and are synthesized in the cytosol, before they are imported into these organelles. The protein traffic between the cytosol and the organelles seems to be unidirectional, since no protein is known to be exported from mitochondria and chloroplasts to the cytosol.

However, during early 1960's it was conclusively shown that mitochondria and chloroplasts also have their own genomes. These organellar genomes encode only some of the organellar proteins, and there is evidence that these proteins encoded by organellar DNA are synthesized within the organelles, utilizing the protein synthesizing machinery available there. There is also evidence that in the remote past, there was horizontal transfer of some genes from mitochondria and chloroplasts to the nucleus, although no proteins encoded by these organelles are exported out of these organelles. Exchange of genetic material also took place between chloroplasts and mitochondria.

The genomes of mitochondria and chloroplasts have been studied in great detail ever since their discovery in 1960's In recent years, mitochondrial genomes of a number of animals and plants as also the chloroplast genomes of a number of plants have been subjected to whole genome sequencing, in view of their small size (see below for details). This information has also been utilized to study the origin of mitochondria from a class of α-proteobacteria and that of chloroplasts from some members of Cyanophyceae (blue green algae). In this chapter, we discuss the organization of mitochondrial and chloroplast genomes. These genomes will also be compared with eukaryotic genomes on the one hand and with bacterial genomes on the other hand. It will also be shown, how this information has been used to trace the origin of these genomes from bacterial systems.

Shape, Size and Number of Mitochondrial and Chloroplast Genomes : Almost all eukaryotes have mitochondrial genomes and all photosynthetic eukaryotes (green plants) have chloroplast genomes. With the advent of genomics era, as many as about 500 mitochondrial genomes and as many as about 35 chloroplast genomes have been fully sequenced (sequencing of a genome means determination of nucleotide

Sizes of Mitochondrial and Chloroplast Genomes

Species	*Type of organism*	*Genome size (kb)*
Mitochondrial Genomes		
Plasmodium falciparum	Protozoan (malaria parasite)	6
Chlamydomonas reinhardtii	Green alga	16 (linear)
Mus musculus	Vertebrate (mouse)	16
Homo sapiens	Vertebrate (human)	17
Metridium senile	Invertebrate (sea anemone)	17

Contd...

Species	*Type of organism*	*Genome size (kb)*
Drosophila melanogaster	Invertebrate (fruit fly)	19
Chondrus crispus	Red alga	26
Aspergillus nidulans	Ascomycete fungus	33
Reclinomonas americana	Protozoa	69
Saccharomyces cerevisiae	Yeast	75
Suillus grisellus	Basidiomycete fungus	121
Brassica oleracea	Flowering plant (cabbage)	160
Arabidopsis thaliana	Flowering plant (vetch)	367
Zea mays	Flowering plant (maize)	570
Cucumis melo	Flowering plant (melon)	2500
Chloroplast Genomes		
Pisum salivum	Flowering plant (pea)	120
Marchantia polymorpha	Liverwort	121
Oryza saliva	Flowering plant (rice)	136
Nicotiana tabacum	Flowering plant (tobacco)	156
Chlamydomonas reinhardtii	Green alga	204

sequence of the entire genomic DNA). On the basis of detailed information on so many genomes, we now know that mitochondrial and chloroplast genomes are generally circular in shape, except the mitochondrial genomes of some algae (e.g. *Chlamydomonas),* several yeasts and some protozoa (e.g. *Paramecium),* where the mitochondrial genome is also linear. In some cases, the circular genomes co-exist with the linear genomes.

The genomes of mitochondria and chloroplasts are relatively simple in organization and small in size (Table). The size of mitochondrial genome differs greatly among different organisms, ranging from 6 kb in *Plasmodium* (malarian parasite) to 2500 kb in melon *(Cucumis melo).* The variation in size of mitochondrial genomes is, however, unrelated to the variation in complexity of organisms (Fig.). For instance, most multicellular animals have small mitochondrial genomes with a compact genetic organization, the genes being close together with little space between them. This is exemplified by human mitochondrial genome (Figure), which is only 16,569 bp in length. In contrast, lower eukaryotes such as 5. *cerevisiae* (Figure), as well as flowering plants, have larger and less compact mitochondrial genomes, with a number of genes containing introns. In humans, the two strands of double-stranded mtDNA are also distinguished as heavy strand (H-strand) and light strand (L-strand), since they differ in density, when measured on CsCl density gradient.

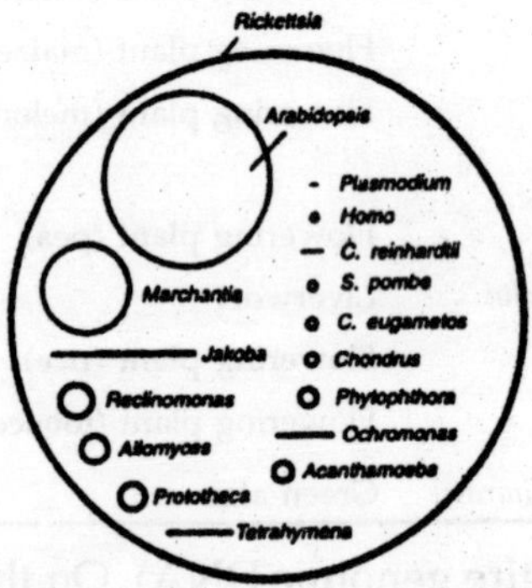

Mitochondrial genome, showing variation in size. For comparison, genome of a mitochondria like proteobacterian of the Rickettsia group (Reclinomonas americana) is also shown.

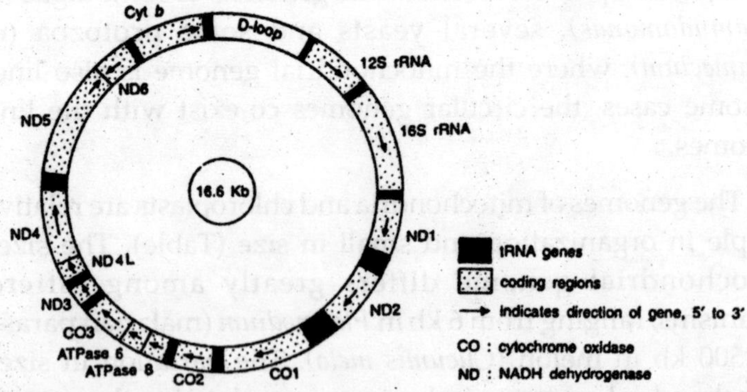

The human mitochondrial genome showing (i) 2 rRNA genes, (ii) 13 protein coding regions and (iii) regions containing genes for 22 tRNAs (14 transcribed in the clockwise direction and 8 transcribed in the anti-clockwise direction).

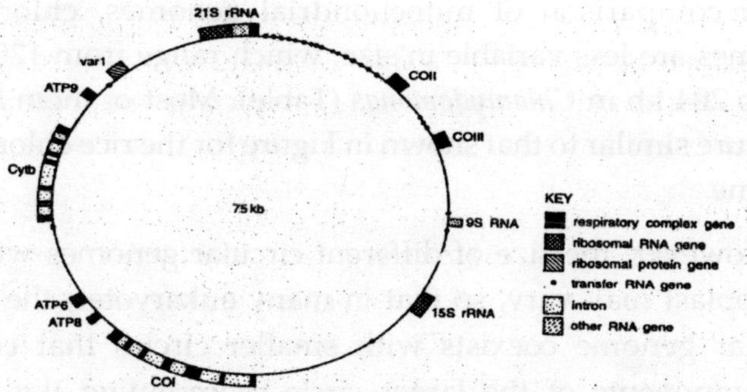

The Saccharomyces cerevisiae (budding yeast) mitochondrial genome. In the yeast genome, the genes are more spaced out than in the human mitochondrial genome and some of the genes have introns. This type of organization is typical of many lower eukaryotes and plants. (Abbreviations: ATP6, ATP8, ATP9, genes for ATPase subunits 6, 8 and 9, respectively: COI, COII, COIII, genes for cytochrome c oxidase subunits I, II and III respectively; Cyt b, gene for apocytochrome b; var I, gene for a ribosome-associated protein. Ribosomal RNA and transfer RNA are two types of non-coding RNA. The 9S RNA gene specifies the RNA component of the enzyme ribonuclease P.)

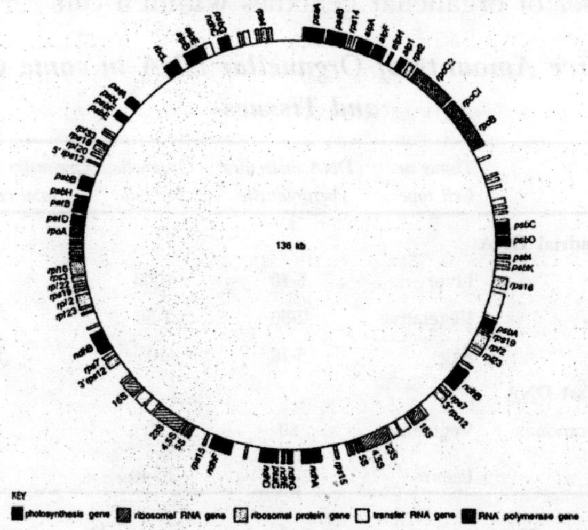

The rice chloroplast genome. Only genes with known functions are shown. A number of genes contain introns which are not indicated on this map. These discontinuous genes include several of those for tRNAs; this is why the tRNA genes are of different lengths even though the tRNAs that they specify are all of similar size.

In comparison of mitochondrial genomes, chloroplast genomes are less variable in size, which range from 120 kb in pea to 204 kb in *Chlamydomonas* (Table). Most of them have a structure similar to that shown in Figure for the rice chloroplast genome.

However, the size of different circular genomes within a chloroplast may vary, so that in many eukaryotes, the larger circular genome coexists with smaller circles that contain subcomponents of the larger circle representing the whole genome of the chloroplast. In an extreme situation available in dinoflagellates (a group of marine algae), chloroplast genomes are split into many small circles, each containing just a single gene.

The number of genomes in an individual organelle also varies, and in turn there is a large number of mitochondria or chloroplasts within an individual cell, thus increasing manyfold the number of organellar genomes within a cell

Relative Amounts of Organellar DNA in some Cells and Tissues

	Organism	*Tissue or Cell type*	*DNA molecules perorganelle*	*Organelle per cell*	*Organelle DNA (as %) total cell DNA)*
I	**Mitochondrial DNA**				
	Rat	Liver	5-10	1000	1
	Yeast	Vegetative	2-50	1-50	15
	Frog	Egg	5-10	10^7	99
II	**Chloroplast DNA**				
	Chlamydomonas	Vegetative	80	1	7
	Maize	Leaves	20-40	20-40	15

(Table). For instance, each human cell contains 800 mitochondria, and each mitochondrion contains as many as 10 identical genomes, so that each cell contains about 8000 mitochondrial genomes.

In yeast, although the number of genomes per mitochondrion is about 100, there are fewer mitochondria per cell, thus making the number of genomes per cell equivalent to that in humans (–5000).

In case of plants, the number of chloroplast genomes per cell is approximately 5000, but that in photosynthetic microorganisms like *Chlamydomonas,* is only 1000 per cell.

The Genetic Content of Mitochondrial and Chloroplast Genomes : Organellar genomes are much smaller than their nuclear counterparts. We therefore anticipate that the gene contents in these organellar genomes are much more limited, which is indeed the case,

Mitochondrial genomes exhibit greater variability in gene content.

Again, mitochondrial genomes display greater variability, the number of genes ranging from 5 for the malaria parasite *(P. falciparum)* to 92 for the protozoan, *Reclinomonas americana* (Table). All mitochondrial genomes contain genes for the non-coding rRNAs and for at least some of the protein components of the respiratory chain, the latter being the main biochemical feature of the mitochondrion.

The more gene-rich mitochondrial genomes also code for tRNAs, ribosomal proteins and proteins involved in transcription and translation; they also code for proteins involved in transport of other proteins into the mitochondrion from the surrounding cytoplasm (Table, Fig). The organization of the human mitochondrial genome is shown in Figure.

Features of Mitochondrial Genomes

Feature/Types of genes	*Plasmodium falciparum*	*Chlamydomonas reinhardii*	*Homo sapiens*	*Saccharomyces cerevisiae*	*Arabidopsis thaliana*	*Reclinomonas americana*
Genome size (kb)	6	16	17	75	367	69
Total number of genes	5	12	37	35	52	92
Protein-coding genes	5	7	13	8	27	62
Respiratory complex	0	7	13	7	17	24
Ribosomal proteins	0	0	0	1	7	27
Transport proteins	0	0	0	0	3	6
RNA polymerase	0	0	0	0	0	4
Translation factor	0	0	0	0	0	1
Non-coding RNA genes	0	5	24	26	25	30
Ribosomal RNA genes	0	2	2	2	3	3
Transfer RNA genes	0	3	22	24	22	26
Other RNA genes	0	0	0	1	0	1
Number of introns	0	1	0	8	23	1

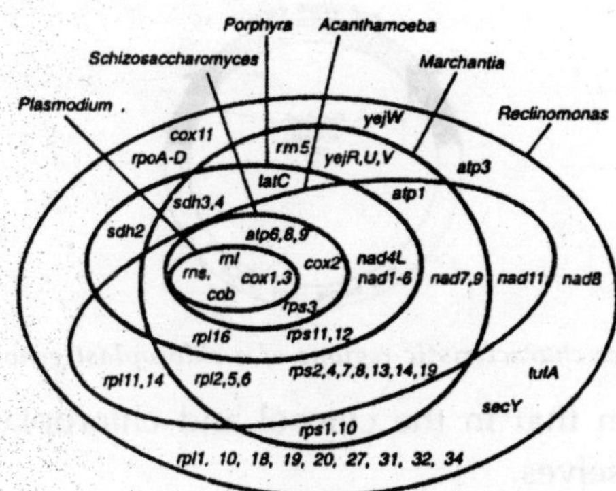

Mitochondril genomes of a variety of organisms showing variation in gene content.

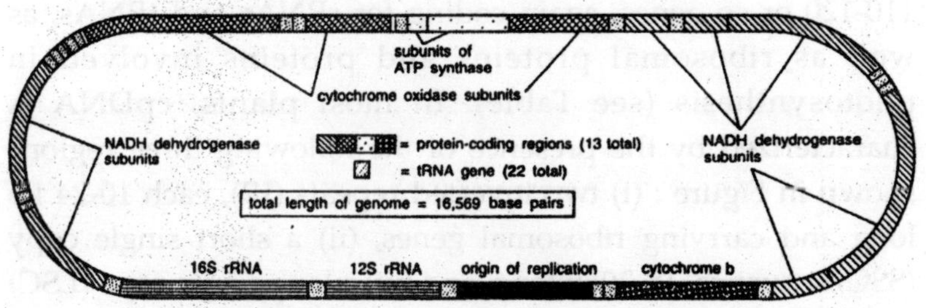

The Organization of human mitochondril genome.

Why are Mitochondrial Genomes Much Larger in Plants? : Mitochondrial genomes are much larger in plants (150 kb to 2500 kb) than in animals (6 kb to 70 kb). It has been shown that perhaps much extra DNA in plant mitochondria is 'junk DNA'.

Mitrochondrial Genome has Several Surprising Features : The mitochondrial genome also has several surprising features, which include the following : (i) there are very few regulatory DNA sequences, most sequences making a part of one or the other coding sequence; (ii) only 22-26 tRNA genes encoding 22 tRNAs are found in mitochondria as against 30 tRNAs found each in the cytosol and the chloroplasts; (iii) for at least four of the 64 codons, the genetic code in mitochondria differs

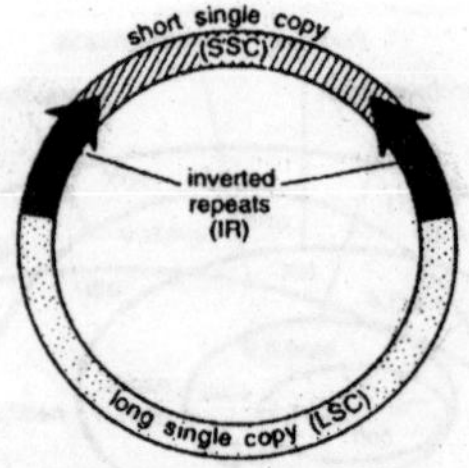

Three characteristic regions of a chloroplast genome

not only from that in the cytosol and chloroplasts, but also among themselves.

Most Chloroplasts Possess Same Set of Genes : Most chloroplast genomes appear to possess the same set of about 110-120 or so genes, again coding for rRNAs and tRNAs, as well as ribosomal proteins and proteins involved in photosynthesis (see Table). In most plants, cpDNA is characterized by the presence of the following three regions shown in Figure : (i) two inverted repeats (IR), each 10-24 kb long and carrying ribosomal genes, (ii) a short single copy (SSC) sequence, 18-20kb long, and (iii) a long single copy (LSC) sequence. About 110 genes that are encoded in cpDNA are listed in Table.

Some Organelle Genes Contain Introns : Unlike human mtDNA, some plant and fungal mitochondrial genes contain introns, which need to be spliced out from RNA transcripts during processing. Introns have also been found in at least 20 chloroplast genes. The presence of introns in these genes is surprising because these are not common in bacteria, which are assumed to have given rise to mitochondria and chloroplasts.

In yeasts, the same mitochondrial gene may have an intron in one strain, but not in the other strains. Such introns perhaps behave like transposable elements (TEs). In some cases, introns were located at the same position in the same gene in

Genes Encoded in cpDNA

RNA/Protein	*Genes*	*RNA/Proteins*	*Genes*
1. Genes for protein synthesis apparatus		**2. Genes for photosynthesis apparatus**	
16S rRNA	1	Photosystem I	2
23S rRNA	1	Photosystem II	7
4.5S rRNA	1	Cytohdrome b/f	3
5S rRNA	1	Ht-ATPase	6
tRNA	30	NADH dihydrogenase	6
r-proteins	19	Ferredoxin	3
RNA polymerase	3	RUBP carboxylase	1
		3. Unidentified genes	**26**
		Total	**110**

the mitochondrial genomes of yeast, *Aspergillus* and *Neurospora*, suggesting their common origin. These observations also suggest that introns have ancient origin and were lost in bacterial genomes and retained in some organellar and nuclear genomes of eukaryotes.

Organellar Protein Synthesis and their Quasi-autonomous Feature : A general feature of organellar genomes emerges from Tables 9.2 to 9.4. These genomes specify some of the proteins found in the organelle, but not all of them. The other proteins are coded by nuclear genes, synthesized in the cytoplasm, and transported into the organelle. If the cell has mechanisms for transporting proteins into mitochondria and chloroplasts, then why not have all the organelle proteins specified by the nuclear genome? We do not yet have a convincing answer to this question, although it has been suggested that at least some of the proteins coded by organellar genomes are extremely hydrophobic and cannot be transported through the membranes that surround mitochondria and chloroplasts, and so simply cannot be moved into the organelle from the cytoplasm. The only way in which the cell can get them into the organelle is to make them there itself.

Mitochondria and Chloroplasts Resemble Prokaryotes in Protein Synthesis : It is known now that both mitochondria and chloroplasts contain the four components necessary for autonomy, that is, DNA, DNA polymerase, RNA polymerase and a protein synthesizing system. Most of these components and the mechanism of protein synthesis in mitochondria and chloroplasts resemble those in prokaryotes. For instance, the RNA polymerase in mitochondria resembles that in certain bacteriophage (single subunit), and RNA polymerase in chloroplasts largely resembles that in *E. coli.* Also the mRNA in both these organelles lack a 5' cap (5' cap is characteristic of all eukaryotic mRNAs) and is synthesized as polycystonic mRNA (as in *E. coli)*

Boths Strands of mtDNA are Transcribed : In human cells, both strands of mtDNA are transcribed at the same rate from a single promoter region on each strand, but in opposite directions. Thus two different giant RNA molecules are produced. The transcripts made on heavy strand (H-strand), are extensively processed to yield two rRNAs, most tRNAs and 10 poly-A containing mRNAs (no 5' cap is found). The transcripts from the light strand (L-strand) is processed to produce only 8 tRNAs and one small poly-A containing mRNA (remaining 90% of L-strand transcript is degraded)

Mitochondria and chloroplasts both contain 70S ribosomes resembling those of bacteria but differing from 80S ribosomes found in cytoplasm. In leaves of higher plants, chloroplast ribosomes make 50% of the total cellular ribosomes. The mitochondrial and chloroplast ribosomes resemble those in prokaryotes in following additional features (i) use of N-formyl methionine as initiating amino acid, (ii) sensitivity to chloramphenicol, lincomycin and spectinomycin which inhibit protein synthesis, and (iii) basic structures of their RNA and proteins.

The role of chloroplast and mitochondrial ribosomes in protein synthesis has been demonstrated in the past through selective inhibition of protein synthesis in these organelles by chloramphenicol. A number of mutants deficient in some organellar proteins but segregating in Mendelian fashion also indicated that at least some organellar proteins are under nuclear control. There were other mutants showing maternal inheritance thus indicating their location in chloroplasts.

Mitochondrial Diseases

During the last 15 years (1990 onwards), several mitochondrial defects have been found to be associated with diseases in human beings and mouse (used as a model). About 50 pathogenic mtDNA base substitutions and rearrangements (deletions and insertions) have been identified in a variety of degenerative diseases. Other diseases involved defects in mitochondrial bioenergetics or biogenesis, which are linked with mutations in the nuclear genes. The three important functions of mitochondria (covered by oxidative phosphorylation), defects in which are related with diseases include the following : (i) energy production, (ii) generation of reactive oxygen species (ROS) and (iii) regulation of programmed cell death or apoptosis.

Gradual Emergence

Monophyletic Origin of Mitochondrion from a Free Living Proteobacterium : Mitochondria resemble bacteria in many respects (Table) and are believed to have originated from some

Comparison between Mammalian Mitochondria and Bacteria

Characters	*Mitochondria*	*Bacteria*
1. Size : width	0.2— 1.0 μ	0.2—2 μ
length	1.0—4 μ	0.3—10.0 μ

Contd...

Characters	*Mitochondria*	*Bacteria*
2. Lipoprotein membrane	6—7 (nm)	7—8 (nm)
3. Invaginations of membrane	Cristae	Mesosomes
4. Respiratory control	Marked	Low or absent
5. Inhibitors of phosphorylation e.g. CN—, CO, azide, antimycin, DNP and oligomycin)	Effective	Less effective or ineffective
6. DNA shape	Closed circle	Closed circle
7. Inhibitor of protein synthesis	Chloramphenicol	Chloramphenicol
8. Ribosomes	70S (variable)	70S

bacteria, rickettsial subdivision of the α -proteobacteria. During 1997-99, complete DNA sequences of bacteria like mitochondrial genome (e.g. *Reclinomonas americana;* 69,034 bp) and mitochondria like eubacterial genome (e.g. *Rickettsia prowazekii;* 1,111,523 bp) were worked out, suggesting strongly that all extant mtDNAs had their origin in a single ancestral protomitochondrial genome, which in its turn, must have originated from a eubacterial genome (monophyletic origin rather than polyphyletic origin).

A remarkable variation exists in the size of mtDNA and its gene content (Figs.). Based on this information, mtDNA has been classified in two basic types : (i) ancestral type with many genes as in *R. americana* and (ii) derived type marked with extensive gene loss, divergence in coding and non-coding DNA sequences, a new mitochondrial code, RNA editing, etc. (e.g. animal and fungal mtDNAs).

Contemporenous Emergence of Eukaryotic Cells and Mitochondria ('Serial Endosymbiosis' vs 'Hydrogen Hypothesis') : A serial endosymbiosis model has also been proposed for the origin of nuclear genome and mitochondria. In this model, it is assumed that in the first step, the nuclear genome of the host itself resulted from the fusion of eubacterial and archaebacterial partners and that only in the second step, the mitochondrion was acquired as a symbiont. A group of eukaryotes known as Archezoa, do lack mitochondria, suggesting that perhaps eukaryotes without mitochondria originated first as a primitive form and that the mitochondria entered the cells later.

The above view, however, has recently (1997-99) been challenged and it is suggested that the eukaryotes acquired the nucleus and mitochondria simultaneously, and that Archezoa (lacking mitochondria) must have resulted due to loss of mitochondria. This view was supported by the observation that genes for several mitochondrial proteins of bacterial origin are found in the genome of amitochondriate Archezoa and also in the organelle hydrogenosome, recently discovered in protists lacking mitochondria. This view is described as 'hydrogen

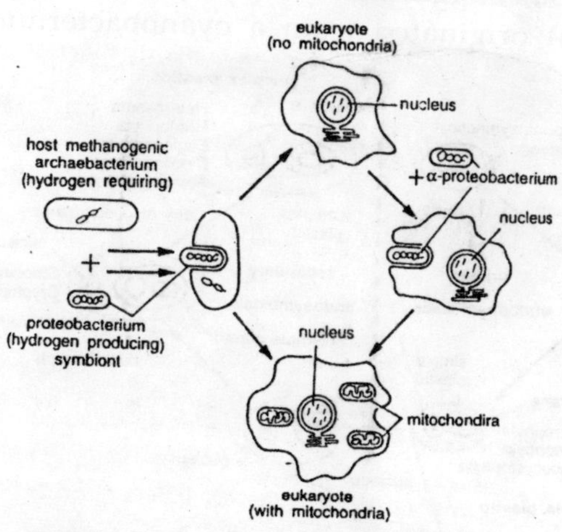

Two alternative hypotheses for the origin of eukaryotic cell and its mitochondrion.

hypothesis' since it assumes chimeric origin of eukaryotic nucleus, due to symbiotic association between (i) a eubacterium (a proteobacterium, the symbiont), producing H_2 as the end product of anaerobic metabolism and (ii) a hydrogen - requiring autotrophic archaebacterium (the host). This hypothesis thus allows the possibility of simultaneous origin of the ancestor of eukaryotic cell and its mitochondrion (Fig.).

Coincident Birth of Plastids from Cyano-bacteria and "Endosymbiotic Gene Transfer" : We briefly discussed the monophyletic endosymbiotic origin of mitochondria from proteobacteria. During 1990s, due to the availability of extensive data, on the nucleotide sequences of a number of chloroplast genomes and the corresponding functional genes in each case, it became possible to apply the principle of comparative genomics in case of chloroplasts also. One of the smallest genomes of cyanobacteria, sequenced in the recent past is 3,573,470 bp long and belongs to *Synechocystis;* in contrast, the biggest chloroplast genome sequenced so far (till 1998) is 191,028 bp long, which belongs to *Porphyra. A* comparison of sequences of chloroplast genomes from 12 species with those of cyanobacteria, suggested that the plastids (particularly chloroplasts) originated from a cyanobacterium, due to

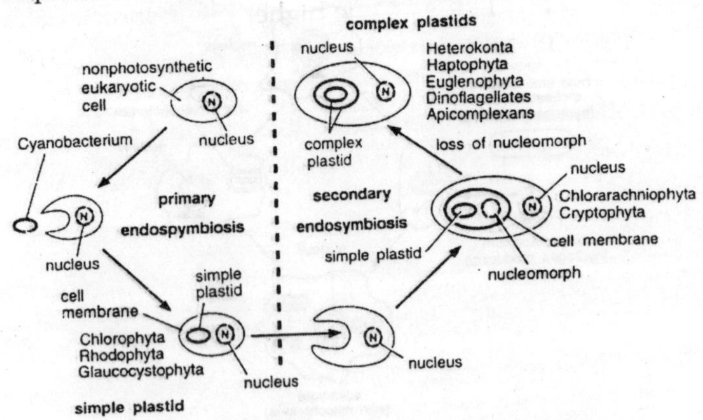

A model for the evolutionary origin of simple and complex plastids from a cyanobacterium through endosymbiosis.

endosymbiosis followed by extensive 'endosymbiotic gene transfer' from the cyanobacterium to the eukaryotic host mucleus. While a primary endosymbiosis led to the origin of simple two-membraned plastids, a secondary endosymbiosis is assumed to be involved in the origin of complex three-or four-membraned plastids (Fig.). As in case of mitochondria, it is also believed that during evolution, chloroplasts exported their genes to the nucleus, but reimported the products with the help of transit peptides and protein import machinery. Sufficient evidence is now available to suggest that endosymbiotic gene transfer' from organelles to nucleus really took place in eukaryotic cells.

It has also been shown that the plastid genome though generally encodes many more (21 to 200) proteins, relative to mitochondrial genome (5 to 63), but they still code only 1% to 5% of the protein-coding genes found in a relatively small cyanobacterium genome. In contrast to this small number of genes, the proteins that are found in a fully functional plastid have been estimated to be up to 5000 in *Arabidopsis*, suggesting that vast majority of these proteins are imported. Recent analysis also suggested that a minimum of 50 genes, that are found in at least one plastid genome, have functional homology with the nuclear genes of at least one higher plant, indicating large scale transfer of genes from plastid to the nucleus. There are also proteins in the chloroplasts, which are encoded in the nuclear genes, but perhaps originally belonged to mtDNA, so that some of the chloroplast proteins were originally encoded in the mtDNA.

5

Chemistry of the Gene

Of the two main types of nucleic acids (DNA and RNA), DNA is known to be the genetic material in most eukaryotes. DNA, therefore, is the most important of all the macromolecules of the cell. All cells of a living organism possess the same quantity and quality of DNA. Since these cells result from a single cell by successive cell divisions, a very precise mechanism for DNA synthesis (replication) should be available in the cell. This DNA is also variously modified to allow for the recognition of (i) old and newly synthesized strands of DNA duplex, (ii) any foreign DNA invading the cell or (iii) the active (expressed) and passive genes. There are also mechanisms for repair of DNA damage in the cell.

In this chapter, details of the mechanism of DNA synthesis (particularly DNA directed DNA synthesis *i.e.* DNA replication, and briefly also RNA directed DNA synthesis i.e. reverse transcription) along with a very brief account of DNA modification (methylation, etc.) and DNA repair will be presented.

DNA Replication : General Features

Semi-conservative DNA replication in *E. coli*

Both DNA as well as genetic RNA are capable of undergoing self-replication. Theoretically, replication of double stranded DNA, could be (i) conservative, (ii) dispersive, or (iii) semi-conservative. The conservative replication would mean that double stranded molecule is conserved as such and that a new copy is synthesized from old molecule. In dispersive replication, the old molecule should disintegrate and two new molecules would be synthesized. In the semi-conservative replication, the two strands would separate from one another, maintain their integrity and each will synthesize, from the pool of nucleotides, its complementary strand. The result would be, that the newly synthesized molecule would carry or conserve one of the two strands from the parent molecule and the other strand would be newly assembled (Fig.). There is sufficient evidence to prove that double stranded DNA really replicates by semi-conservative method.

Meselson and Stahl's experiment (using N^{15}). M. Meselson and F.W. Stahl in 1958 reported the results of an experiment which was designed to test whether double stranded DNA replicates in a semi-conservative manner. The following two basic principles were involved : (i) DNA was labelled with heavy nitrogen (N^{15}) and was then allowed to replicate in a medium containing N^{14}. If replication is semi-conservative, then after first generation of replication, one of the two strands would have heavy nitrogen (N^{15}) and the other strand would have normal nitrogen (N^{14}). The resulting molecule would have a density which is intermediate between N^{15} DNA and N^{14} DNA. This density will gradually fall and in the following generations of replication, density would approach that of N^{14} DNA. (ii) Another principle utilized was the preparation of a caesium chloride density gradient. This density gradient is prepared by gradual dilution of a heavy salt solution. When this is subjected to ultracentrifugation, with a substance having density within the range of this gradient, this substance will find place at its own level of density. This enables detection of very slight differences in density.

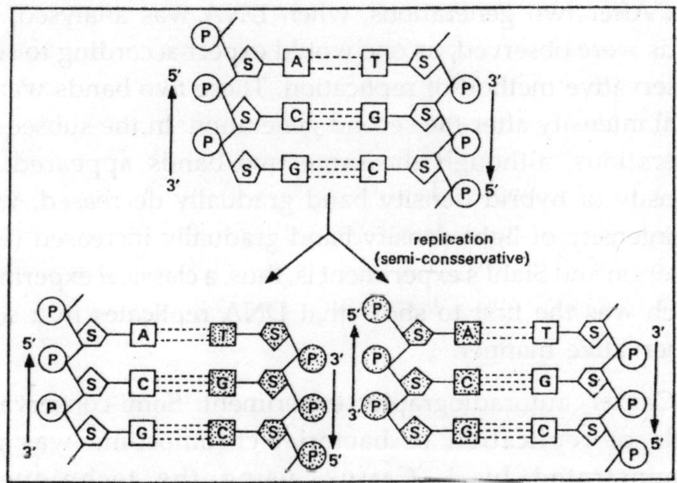

Basic Plan of the structure of double helical DNA and the semi-conservative replication.

Meselson and Stahl (1958) allowed *Escherichia coli* cells to 0grow on N^{15} culture medium for about 14 cell generations, so that almost all nitrogen (N^{14}) in DNA is replaced by N^{15}. Then the cells were abruptly transferred to N^{14} culture medium. Since the time required for one cell generation was determined to be about 30 minutes, it was possible to remove the cells after one or more known number of generations of replication. Their DNA could then be analysed and the following results were obtained.

DNA sample obtained one cell generation after the transfer to N^{14} culture medium showed only one density band as observed through ultraviolet absorption pattern. This band indicated uniform homogeneous density of DNA after one cell generation. Moreover, the band was exactly between the bands formed by N^{15} DNA and N^{14} DNA, indicating that all DNA found after first generation had intermediate density. This is what one would expect if DNA replicates in a semi-conservative manner.

Density of DNA was studied in the following generations also. After two generations, when DNA was analysed, two bands were observed, as one would expect according to semi-conservative method of replication. These two bands were of equal intensity after the second generation. In the subsequent generations, although the same two bands appeared, the intensity of hybrid density band gradually decreased, while the intensity of light density band gradually increased (Fig.). Meselson and Stahl's experiment is, thus, a classical experiment which was the first to show that DNA replicates in a semi-conservative manner.

Cairns' autoradiography experiment. Semi-conservative mode of replication of bacterial chromosome was also demonstrated by J. Cairns, using the technique of autoradiography. In autoradiography technique, the cells are first supplied with a suitable radioactive material like tritiated

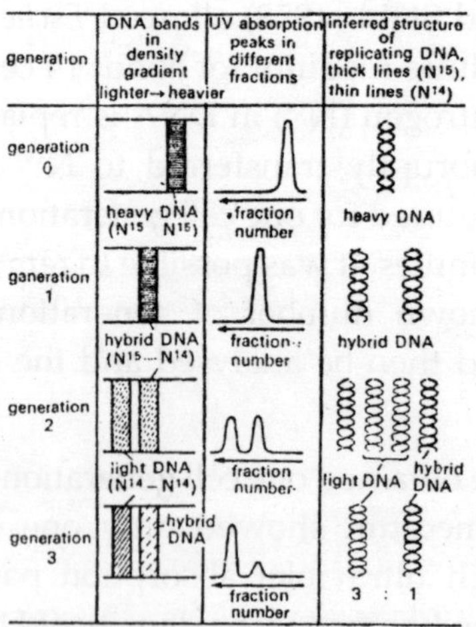

Experiment of Meselson and Stahl demonstrating semi-conservative replication.

thymidine (H —TdR; H^3 is heavy isotope of hydrogen and it replaces normal hydrogen in thymidine to give rise to tritiated thymidine). Tritiated thymidine is used since this will selectively label only DNA and will not label RNA, since thymine base is absent in RNA.

The tritiated thymidine gets incorporated into DNA and replaces ordinary thymidine. The cellular material is then sectioned or else the cells may be broken down to release the intact bacterial chromosomes on slides. These slides are then covered by photographic emulsion and stored in dark. During this storage, the particles emitted by tritiated thymidine will expose the film, which can be developed. This photograph will then show the regions of the presence of tritium and thus indirectly show the presence of labelled DNA.

Using the above technique, replication of DNA could be easily followed by J. Cairns and the results were reported in 1963. During incorporation of tritiated thymidine, autoradiographs could be prepared at regular known intervals. Autoradiographs from this replicating material prepared at

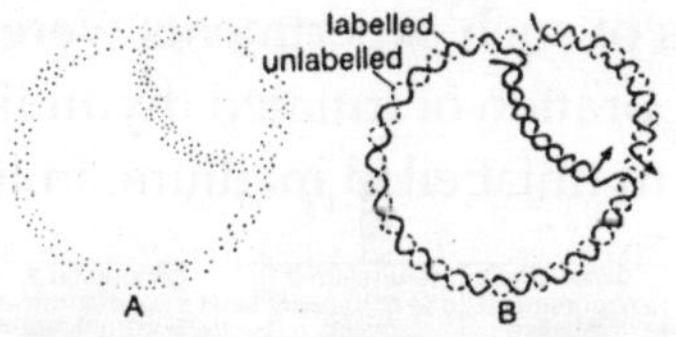

Cairns' autoradiography experiment, showing (A) an autoradiograph (drawing) and,(B) its interpretation in terms of DNA.

regular known intervals demonstrated semi-conservative mode of replication. Lighter density of dots was considered to indicate that only one of the two strands was labelled, while a heavier density of dots indicated that both strands were labelled. Such a situation was actually observed. The rate at which the replication proceeds could also be worked out by meassuring the length of DNA undergoing replication in a known interval of time. The generation time was worked out by Cairns as 30 minutes in *E. coli.* The length of the chromosome was worked out to about 1 mm. The rate of replication thus would be approximately 30 μ -40 μ per minute (1 mm = 1000 μ).

Figure indicates that after replication, one of the two strands in the daughter DNA molecules is derived from the parent molecule and the other is synthesized afresh. In θ shaped figure, which is obtained in the second cycle of replication in presence of label, two arcs in the split region would never be equally labelled. For instance, one arc would be twice as heavily labelled as the other arc. This is what was actually observed by Cairns. The observations thus clearly supported the semi-conservative nature of replication. In the above set of two experiments, it was demonstrated that DNA replicates in a semi-conservative manner. Subsequent experiments of different kinds proved the validity of this mechanism.

Semi-conservative Replication of Chromosomes in Eukaryotes

Autoradiography Experiment in Vicia Faba: Autoradiography was also utilized by J.H. Taylor and his co-workers for the study of duplicating chromosomes in the root tip cells of *Vicia faba*. Results of such experiments were first published in 1957. After incorporation of tritiated thymidine, when root tips were transferred to unlabelled medium, in the first generation

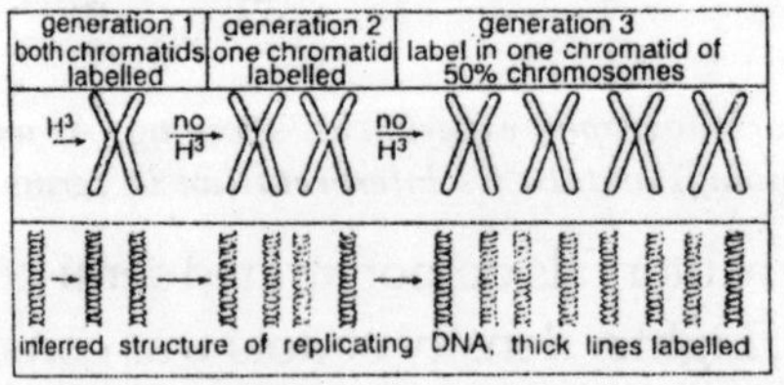

Tayor's experiment on Vicia faba root tips using autoradiography technique.

of duplication both chromatids were labelled (interpreted as one DNA double helix in each chromatid, and only one of the two strands labelled). In the second cycle of duplication, in each chromosome, one of the two chromatids was found to be labelled. This was interpreted as showing semi-conservative mode of duplication (Fig.). Taylor's experiment is important in so far as it demonstrated the semi-conservative mode of replication in chromosomes of a higher plant. But it should not necessarily mean semi-conservative mode of replication of DNA, because the organization of DNA in chromosomes of higher plants and animals is rather complex and not as simple as envisaged by Taylor, while interpreting his results.

Use of harlequin chromosomes. Semi-conservative replication of chromosomes can also be visualized through an examination of chromosomes that are allowed two rounds of replication in a medium containing bromodeoxy-uridine. These replicated chromosomes are then stained with a fluorescent dye and Giemsa stain to produce what are described as *harlequin chromosomes* (Fig.). The newly synthesized DNA stains differently than the parental DNA, so that the harlequin chromosomes have been found suitable for a study of (i) semi-conservative replication, (ii) crossing over and (iii) sister chromatid exchanges.

Semi-discontinuous DNA Replication : A. Kornberg initially characterized an enzyme, now called DNA polymerase I and believed it to be responsible for DNA replication. However, later it was shown that this enzyme is mainly involved in DNA repair and not in DNA replication. Another enzyme DNA

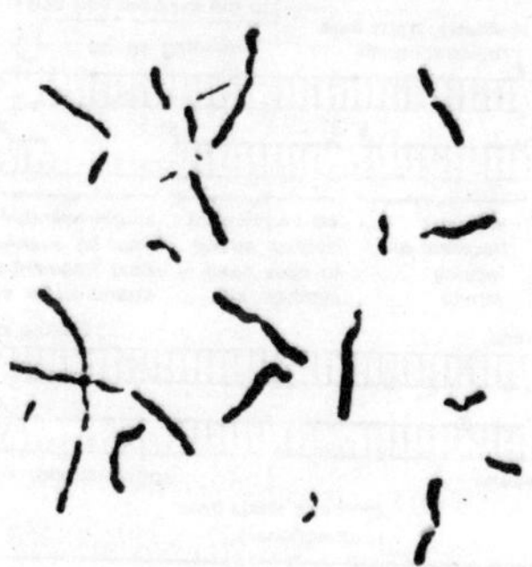

Harlequin chromosomes in a Chinese Hamster Ovary (CHO) cell. Arrow indicates sister chromatids exchange in a chromosome (see text for further details; from S. Wolff and J. Bodycote).

polymerase III is now known to be responsible for DNA replication and synthesizes DNA in 5' → 3' direction. Since the two strands of DNA have opposite polarities, DNA synthesis can not proceed on both strands, utilizing same enzyme, unless the synthesis proceeds in pieces. Such pieces called Okazaki pieces (after the name of discoverer) have actually been observed and it is established that DNA synthesis is discontinuous on one of the two strands and the segments later get fused with the help of ligase enzyme (Fig.).

At one time, it was felt that there was enough evidence to suggest that DNA synthesis is discontinuous on both the strands. However, now it is known that DNA synthesis is continuous on one strand and discontinuous on the other strand (Fig.). When the double helix of DNA unwinds, DNA replication on one of the two strands (3' to 5' strand) can easily proceed continuously in 5' to 3' direction. This is the leading strand. On the other strand (5'—3' strand), if the synthesis has to take place continuously in 5' to 3' direction, it has to be synthesized in a direction opposite to that on the leading strand. This strand is the lagging strand and on this strand, synthesis takes place in segments discontinuously and these segments are then fused to create an intact lagging strand. This

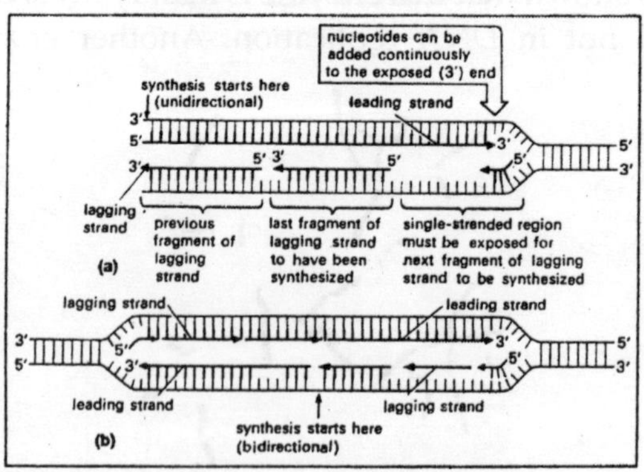

Semi-discontinuous DNA replication under conditions of (A) unidirectional replication, (B) bidirectional replication

behaviour where the leading strand is synthesized continuously and the lagging strand is synthesized discontinuously is called semi-discontinuous replication.

In certain viruses like adenovirus or ϕ 29, linear DNA replicates from the two ends by strand displacement, so that both strands can be copied in 5' to 3' direction simultaneously without any need for discontinuous replication (Fig.). Therefore, it is obvious that semi-discontinuous replication is the result of a need to synthesize both strands simultaneously from the same origin.

Unidirectional and Bidirectional DNA Replication : John Cairns, from his autoradiography experiments which are described earlier in this chapter, had concluded that DNA synthesis starts at a fixed point on the chromosome and proceeds in one direction. It was subsequently realized that Cairns' results could be interpreted in terms of bidirectional replication also. Additional evidences supporting bidirectional replication were obtained from (i) autoradiography, (ii) electron microscopy and (iii) genetic studies. Evidence of bidirectional replication is also available in eukaryotes like yeast, fruitfly and mammals, where DNA synthesis starts at multiple sites,

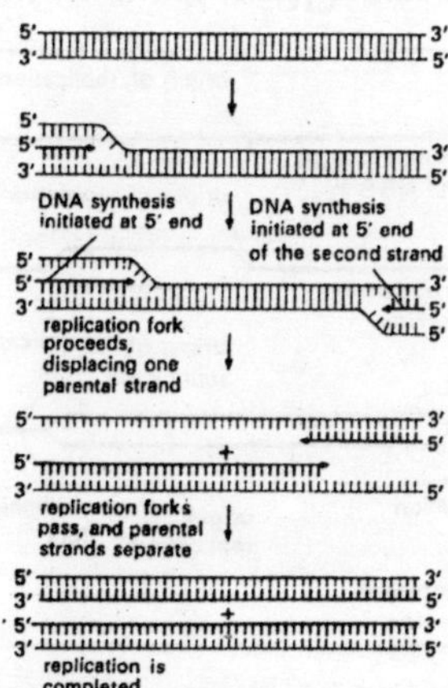

Continuous DNA replication on both strands initiated independently at the two ends (as observed in adenovirus).

which start as loops and can be seen as expanding bubbles or-eyes in electron micrographs (Fig.). Number of eyes within an

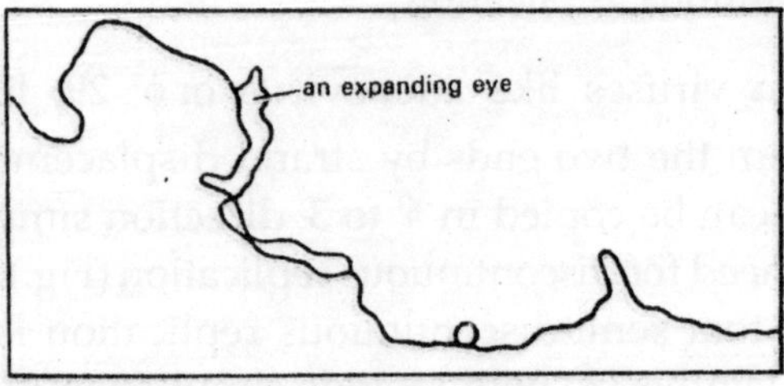

A diagrammatic representation of an electron micrograph of replicating eukaryotic DNA, Showing formation of 'bubble' or 'eye'

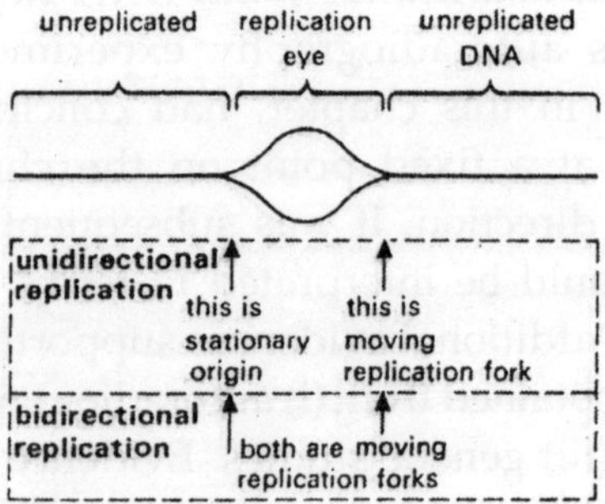

A replication eye may represent unidirectional replication (when only one end moves) or a bidirectional replication (when both ends move).

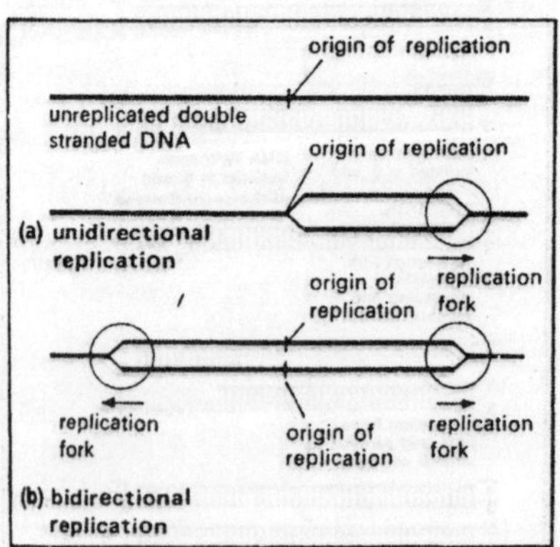

The difference between unidirectional and bidirectional replications in the movement of replication forks.

initiation zone (5000–55000 bp) indicates the number of initiation sites for DNA replication (each about 100-500 bases; see later for details).

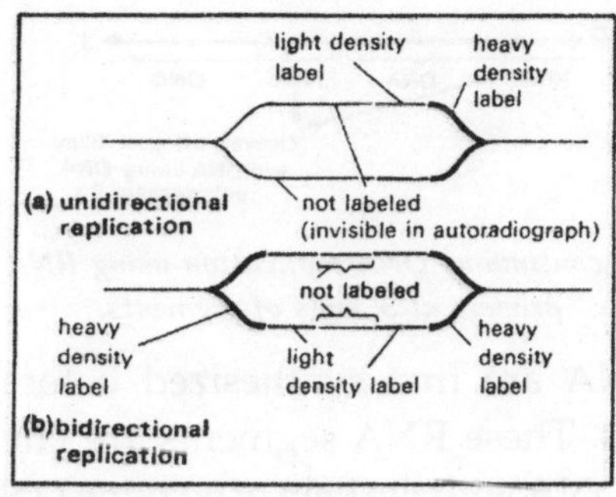

Distinction between unidirectional and bidirectional replications as revealed by radioactive labelling (note increasing densities of labelling on only one side or on both the sides of origin of replication).

In view of the above evidences, it is now known that DNA replication can be unidirectional or bidirectional, depending upon whether the replication from the point of origin proceeds only in one direction or proceeds in both the directions (Fig.). A replication eye may appear in both the situations, unless the replication starts from one of the two ends of a linear DNA molecule. However, in unidirectional replication, one of the two ends of the replication eye will be stationary and the other end will move with replication (Fig.). On the other hand, in bidirectional replication, none of the two ends will be stationary and both will be moving (Fig.). If radioactively labelled nucleotides are used during DNA synthesis, distinction between undirectional and bidirectional replication can be made. This can be done by the study of autoradiographs prepared after DNA synthesis has already proceeded for some time with the use of labelled nucleotides. Labelling on both the replication forks will indicate bidirectional replication and labelling only at one fork will suggest unidirectional replication (Fig.). An example of unidirectional replication is the replication of mitochondrial DNA- (mtDNA) by D-loops in vertebrates.

RNA Primers in DNA Replication : No DNA polymerase enzyme is known, which can initiate synthesis of DNA chain without the availability of a primer strand. Therefore, short

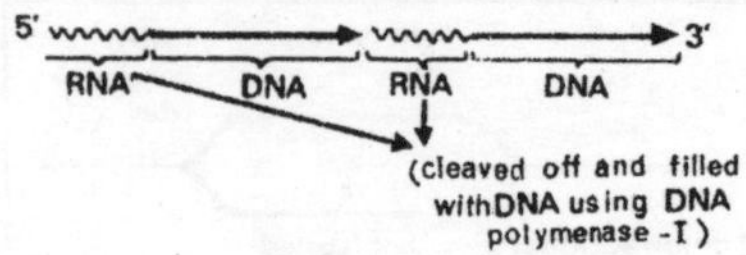

Discontinuous DNA replication using RNA primers at 5' ends of segments.

segments of RNA are first synthesized before actual DNA replication starts. These RNA segments are later excised and gaps filled with DNA with the help of enzyme DNA polymerase I. It was found that if RNA synthesis is stopped by a drug rifampicin, DNA replication stops thus proving the role of RNA primers in DNA replication (Fig.). There are several other evidences which established this feature of DNA replication.

Regulation of DNA Replication by Anti-sense RNA Primers : While studying DNA replication in an *E. coli* plasmid (ColEl), Jun-ichi Tomizawa in USA demonstrated that certain RNA molecules, synthesized on DNA strand complementary to the one that produces RNA primers, are actually synthesized. These were described as antisense inhibitor molecules and the rate of multiplication of the plasmid depended on the ratio of

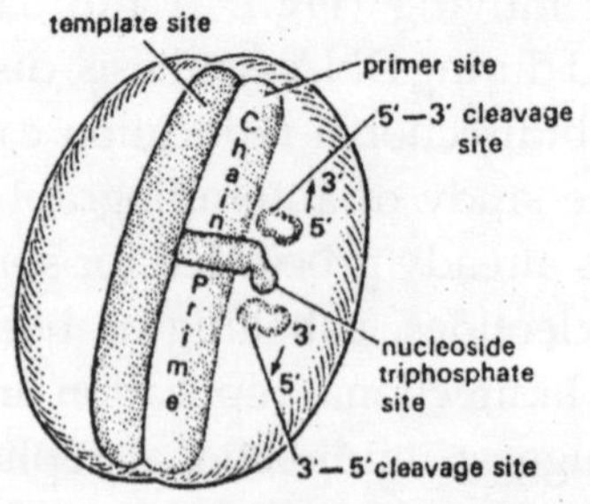

A model of DNA polymerase-I enzyme, showing five different sites. (redrawn from Kornberg, DNA synthesis—1974).

RNA primers to these antisense molecules. Obviously antisense RNA molecules hybridize with RNA primers and render them incapable of initiating DNA replication. Antisense RNA is also used for regulation of protein synthesis (gene expression) and has been shown to offer useful technology for therapeutic uses against certain diseases caused by viruses, etc.

Prokaryotic DNA Polymerases : Three different prokaryotic DNA polymerases are known, of which DNA polymerases I and II are meant for DNA repair and DNA polymerase III is meant for actual DNA replication. (i) DNA polymerase I (isolated around 1960 by Arthur Kornberg) was the first enzyme suggested to be involved in DNA replication. This enzyme is now considered to be a DNA repair enzyme rather than a replication enzyme. It has five active sites shown in Figure and its properties are shown in Table. This enzyme is mainly

A Comparison of Properties of DNA Polymerases of E. coli.

Property	*DNA polymerases*		
	I	*II*	*III*
1. Polymerisation 5' → 3'	yes	yes	yes
2. Exonuclease 3' → 5'	yes	yes	yes
3. Exonuclease 5' → 3'	yes	no	no
4. Use of primed single strands	yes	no	no
5. Use of nicked duplex	yes	no	no
6. Molecular weight	109,000 (single chain)	120,000 (single chain)	>250,000 (heteromultimetric chain)
7. Molecules per cell	400	not known	10-12
8. Gene	*polA*	*polB*	*polC (dnaE), dnattl dnaX, dnuQ, dual*
9. Nucleotides polymerised at 37°, molecules/minute	Up to 1,000	Up to 50	Up to 15,000
10. Affinity for nucleoside triphosphate (NTPs) precursors	Low	Low	High

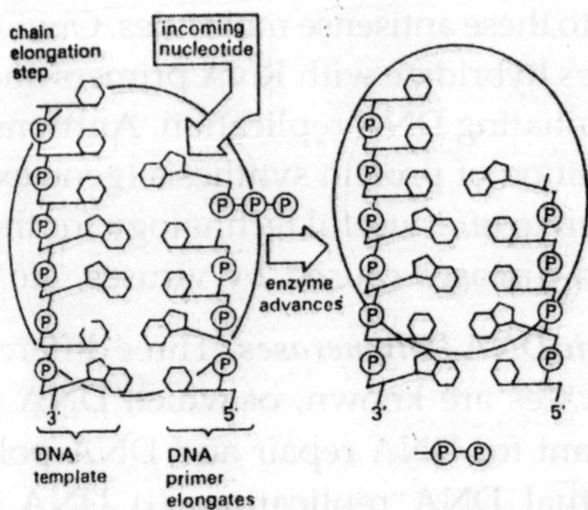

A model showing polymerization of nucleotides, on the primer site of DNA polymerase-I (redrawn from Kornberg, DNA synthesis—1974).

involved in removing RNA primers from Okazaki fragments and fills up the gaps due to its 5'→ 3' polymerising activity. However, a ligase enzyme is needed for covalent bonding after the gap is filled by DNA polymerase I. It can also remove thymine dimers produced due to UV irradiation and fill the gap due to excision. Polymerising activity of DNA polymerase I is shown in Figure and an exonuclease activity removing mispaired nucleotide is shown in Figure. This is described as proofreading function of this enzyme. DNA polymerase I consists of two fragments: a larger fragment, called Klenow fragment (38kD), which contains 3'—5' exonuclease activity with 5'—3' polymerising activity and a smaller fragment (35kD), which contains 5'—3' exonuclease activity.

The activity of this enzyme (due to Klenow fragment) in using nicked DNA *in vitro* is unique and is utilized to label DNA molecules with radioactive nucleotides. This is called nick translation, and is shown in Figure. DNA polymerase I enzyme is synthesized under the control of gene *polA* located on *E. coli* map at a position of 75 minutes. (ii) DNA polymerase II (isolated in 1970-71) resembles DNA polymerase I in its activity to bring about the growth in 5'→ 3' direction, using free 3'—OH groups, but mainly uses duplexes with short gaps only. It can not use nicked duplexes (unlike DNA polymerase I).

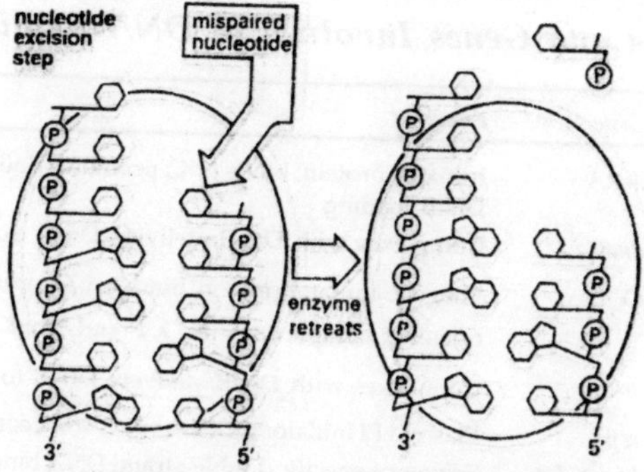

Removal of a mispaired nucleotide by exonuclease activity of DNA polymerase I.

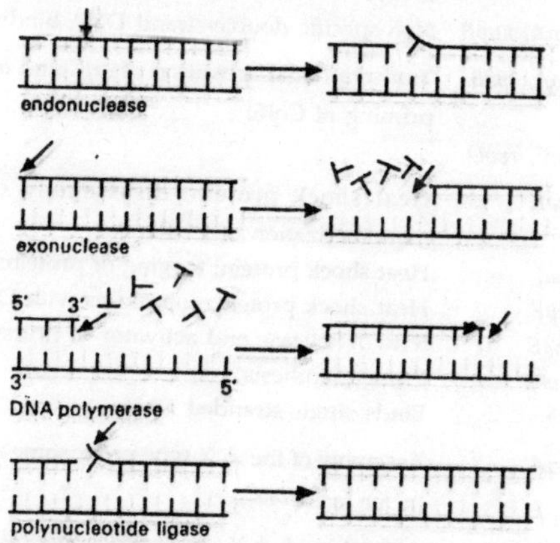

Effect of four classes of enzymes on DNA duplex, showing cleavage in case of nucleases (endonuclease and exonuclease) and repair in the other two cases (redrawn from Watson, Molecular Biology of the Gene—1987).

Although it has 3' → 5' exonuclease activity, it lacks 5' → 3' exonuclease activity (exonuclease activity means cleavage of nucleotides only at the end, while endonuclease breaks DNA strand at an internal position). Since *polB*⁻ mutants lacking DNA polymerase II appear normal in growth and conduct. DNA replication normally, this enzyme can not be a replication enzyme. DNA polymerase II enzyme is also involved in DNA repair, (iii) DNA polymerase III holoenzyme (isolated in 1972)

Proteins and Genes Involved in DNA Replication

Protein	*Gene(s)*	*Function*		
DnaA	*dnaA*	Initiator protein; binds *oriC;* promotes double helix opening; DnaB loading		
DnaC	*dnaC*	Complexes with DnaB; delivers DnaB to DNA		
λ *O*	O	Phage λ initiator protein; binds ori λ ; promotes double helix opening; complexes with λ *P* and DnaK		
λ *P*	*P*	Complexes with DnaB; delivers DnaB to ori λ complex		
PI RepA	*repA*	Plasmid PI initiator protein; binds *oriPI;* complexes with DnaJ.		
IHF	*himA, hip*	Sequence-specific double-strand DNA binding protein; binds to *oriC*		
FIS	*fis*	Sequence-specific double-strand DNA binding protein; binds to *oriC*		
HU	*hupA, hupB*	Non-specific double-strand DNA binding protein		
RNA polymerase	*rpoA, rpoB,*	Transcriptional activation of *oriC* and *ori* λ, leading strand priming of ColEl		
	rpoC, rpoD			
DnaK	*dnaK*	Heat shock protein; disassembly of ori λ, complex; monomerization of P1 RepA		
DnaJ	*dnaJ*	Heat shock protein; tagging of proteins for action by DnaK		
GrpE	*grpE*	Heat shock protein; complexes with DnaK		
DnaB	*dnaB*	5' to 3' helicase and activator of primase		
Primase	*dnaG*	Primer synthesis		
SSB	*ssb*	Binds single-stranded DNA		
PriA	*priA*	Assembly of the ϕ X-type primosome at *pas;* 3' to 5' helicase; (n′ or factor Y)		
PriC	*priC*	Assembly of ϕ X-type primosome (*n*″ protein)		
DnaT	*dnaT*	Assembly of ϕ X—type primosome (protein i or factor X)		
DNA polymerase III holoenzyme				
α	*dnaE*	Polymerase		
ε	*dnaQ*	3' to 5' exonuclease	Core	
θ	*holE*	Stimulates ε	Pol III′	
τ	*dnaX*	Dimerization of core, ATPase		
γ	*dnaX*	Binds ATP		Pol III″
δ	holA	Binds to β		
δ'	*holB*	Cofactor for γ ATPase and stimulates clamp loading, γ complex′		
χ	*holC*	Binds SSB		

Contd...

Protein	*Gene(s)*	*Function*
ψ	*holD*	Bridge between χ and γ
β	*dnaN*	Processivity factor; 'sliding clamp'
PolI	*poiA*	Primer removal and gap filling; initial leading strand synthesis of ColEI
RNase H	*rnh*	Primer removal, primer processing for leading strand synthesis of ColEI
Ligase	*lig*	Joins nascent DNA fragments
Gyrase	*gyrA, gyrB*	Type II topoisomerase; replication swivel, DNA supercoiling, decatenation
Topo IV	'*parC, parE*	Decatenation; presumptive type II topoisomerase
TBP	*tus*	Ter binding protein; stalling of replication forks

plays an essential role in DNA replication. There is intracellular scarcity of this enzyme (only 10-20 molecules per cell), thus requiring rapid recycling. It is a heteromultimeric enzyme with ten units. All the ten subunits, listed in Table, are needed for DNA replication *in vitro*. The ten subunits of holoenzyme are constituted into the following four components: (i) a polymerase 'core' enzyme consisting of one molecule each of subunits α, ε and θ, and having the ability to synthesize DNA; (ii) a sliding clamp, consisting of a dimer (β_2), which, binds to DNA in an ATP dependent reaction catalyzed by 'γ complex' to form the pre-initiation complex (PIC); (iii) 'γ complex' consisting of six polypeptides with five subunits $(\gamma_2\delta_1\delta'_1\chi_1\psi_1)$ it is also called matchmaker, because it helps in the loading of sliding clamp (β_2) onto the DNA. (iv) the τ subunit (used as a dimer, τ_2), which is a DNA dependent ATPase of ill defined function. During initiation of DNA synthesis, sliding clamp is first loaded onto DNA in an ATP dependent reaction catalyzed by matchmaker 'γ complex' to form the pre-initiation complex. This is followed by the assembly of the core enzyme with the pre-initiation complex to form the initiation complex in an ATP independent reaction, during which, the matchmaker (γ complex) dissociates (Fig.). The sliding clamp confers processivity (tendency to remain on a single template rather than to dissociate and reassociate again) to core enzyme and is passively pulled along with core during polymerization. The slilding clamp (β_2) is a ring shaped, head to tail, dimer with a central cavity of sufficient diameter to accommodate duplex DNA. It is a homologue of eukaryotic PCNA (see later).

DNA pol holoenzyme (HE) has an asymmetric structure, consisting of 16 polypeptides involving 10 subunits as follows : $\beta_2(\alpha\varepsilon\theta)_2\,\tau_2(\gamma_2\delta_1\delta'_1\,\chi_1\psi_1)$. The subunits of holoenzyme assemble in an orderly fashion as shown in Figure. Following sub-assemblies of the holoenzyme have also been recognized : (i) Pol III′ consists of two cores and a dimer of $\tau(\tau_2)$, so that the polymerase acts in pair for coordinated replication of both strands of a duplex. (ii) Pol III assembly contains 14 polypeptides $\{(\alpha\varepsilon\theta)_2\,\tau_2(\gamma_2\delta_1\delta'_1\,\chi_1\psi_1)\}$ involving 9 subunits and lacks the sliding clamp or β_2, which confers high speed and processivity to polymerases. Table shows a comparison of

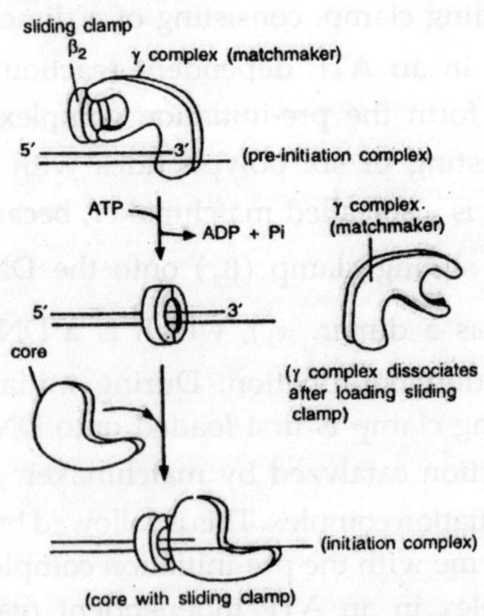

Formation of 'initiation complex' for DNA replication by the assembly of core enzyme (not the role of 'sliding clamp

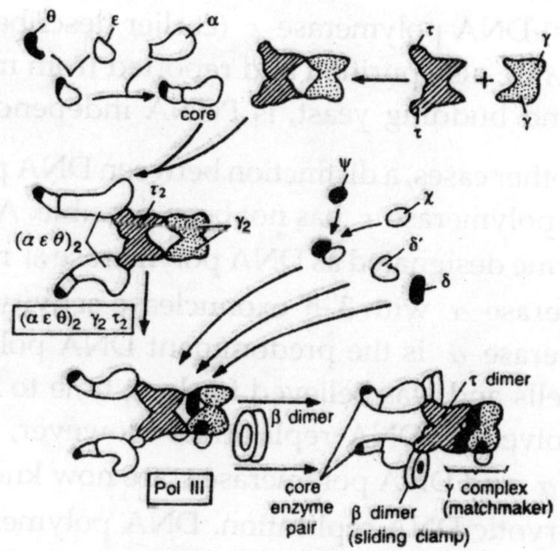

Assembly of DNA polymerase holoenzyme (HE) from its 16 subunits.

DNA polymerase *and* 'γ *complex').* III with DNA polymerase I and DNA polymerase II of *E. coli.* A comparison of synthetic activity of DNA polymerases with other related enzymatic activities (endonuclease, exonuclease, ligase) is presented in Figure.

Eukaryotic DNA Polymerases : Five different eukaryotic DNA polymerases have been identified, from materials like yeast cells, rat liver, human cell cultures and tumour cells. These include the following : (i) DNA polymerase α , a relatively high molecular weight enzyme, also called *cytoplasmic polymerase* (also appropriately called *large polymerase,* since it is also found in the nucleus and since another small nuclear polymerase is also available in cytoplasm); (ii) DNA polymerase β or *nuclear polymerase* (also called *small polymerase),* found only in vertebrates, and not in budding yeast, nor in other lower eukaryotes; (iii) DNA polymerase γ or *mitochondrial polymerase,* encoded in the nucleus; (iv) DNA polymerase δ , another novel enzyme reported from mammalian cells, is PCNA dependent for DNA-synthetic processivity (PCNA = proliferating cell nuclear antigen; processivity = property of continuous synthesis); (v) DNA polymerase ε (earlier described as DNA polymerase δ II), also purified and reported from mammalian HeLa cells and budding yeast, is PCNA independent.

In some other cases, a distinction between DNA polymerase δ and DNA polymerase ε has not been possible. Another calf thymus enzyme designated as DNA polymerase δI represented DNA polymerase α with 3'-5' exonuclease activity. The large DNA polymerase α is the predominant DNA polymerase in eukaryotic cells and was believed for long time to be the only enzyme involved in DNA replication. However, both DNA polymerase α and DNA polymerase δ are now known to take part in eukaryotic DNA replication. DNA polymerase α has four subunits and an appropriate designation for this enzyme is DNA polymerase α /primase. Three other enzymes which are found associated with large DNA polymerase α include *thymidine kinase, ribonucleotide reductase* and *thymidylate synthetase.* They are also believed to be involved in DNA replication.

Nuclear small DNA polymerase β resembles large DNA polymerase α in its replicative activity, but differs in some other respects. Its quantity remains unchanged, while that of large polymerase changes with need. It is speculated that small enzyme may be involved in repair and recombination. The differences between these polymerases are shown in Table.

Eukaryotic DNA Polymerases : Protein Structure, Cellular Location and Properties (Wang, 1991).

	DNA polymerase				
	α *(Pol I)*	β	γ	δ *(Pol II)*	ε *(Pol III)*
Protein structure					
catalytic poly-peptide (kDa)	165[a]	40	140	125	255

Contd....

	DNA polymerase				
	α (*Pol I*)	β	γ	δ (*Pol II*)	ε (*Pol III*)
associated sub-units (kDa)	70 58 48	None	Un-known	48	Unknown
Yeast gene	*POLI (CDC17)*	None	*MIP1*	*POL3 (CDC2)*	*POL2*
Cellular location	Nuclear	Nuclear	Mito	Nuclear	Nuclear
Associated enzyme activities					
5'-3' exonuclease	None	None	None	None	None
3'-5' exonuclease	None[b]	None	Yes	Yes	Yes
primase	Yes	None	None	None	None
Properties					
processivity					
inherent	Med	Low	High	Low	High
with PCNA	—	—	—	High	—
fidelity	High	Low	High	High	High
Inhibitors					
aphidicolin	Yes	No	No	Yes	Yes
N-ethylmaleimide	Yes	No	Yes	Yes	Unknown
butylphenyl-dGTP	Yes	No	No	No	No
dideoxynucleoside 5'-triphosphate (ddNTP)	No	No	Med	No	No
carbonyl-diphosphonate	No	Un-known	Un-known	Yes	Med

* *molecular mass deduced from primary protein sequence,*

* *cryptic exonuclease in the isolated catalytic polypeptide was reported in Drosophila melanoguster embryo.*

Replicons for DNA Replication : DNA replication in prokaryotes as well as in eukaryotes is achieved in discrete units called replicons, which may vary in a genome from one in bacteria like *E. coli,* and 500 in yeast to several thousands in animals and plants (Table). In *E. coli* there is a single replicon with the origin, identified as a genetic locus *oriC* (245 bp). The sequence of this locus has been isolated and cloned in plasmids and allows autonomous replication of plasmid DNA just like bacterial chromosome. Isolated and cloned telomeres and centromeres from yeast are also similarly used for preparing yeast artificial chromosomes (YAC's) The origins of replication have been identified in bacteria, yeast, chloroplasts and mitochondria. These are A:T rich, a feature which is related to unwinding of DNA to initiate replication. In *E. coli* there are also termination sites *(ter A-F),* each consisting of ~ 23 bp. Termination of DNA replication requires product of *tus* gene (*Tus* protein or ter binding protein = TBP), which recognizes *ter* sites.

In a eukaryotic genome, there are many replicons, an individual replicon being small and the rate of replication being much slower. Not all replicons start replication at the same time, but the initiation of replication at different replicons takes place in an ordered manner. There is no evidence of termination sites in eukaryotes (unlike *E. coli),* but termination is brought about when a termination fork meets another fork proceeding towards it from origin of an adjoining replicon.

Prokaryotic and Eukaryotic Replicons

Organism	*No. of replicons*	*Average length (kb)*	*Fork movement (bp/min)*
Bacteria *(E. coli)*	1	4200	50,000
Yeast *(S. cerevisiae)*	500	40	3,600
Fruitfly (D. *melanogaster)*	3500	40	2,600
Toad *(X. laevis)*	15000	200	500
Mouse *(M. muxculus)*	25000	150	2,200
Bean *(Vicia faba)*	35000	300	not known

DNA Replication in Prokaryotes

Among prokaryotes, DNA replication *in vitro* has been studied in great detail in *E. coli* and its phages and plasmids. The techniques of genetics and biochemistry have been used to identify genes and their products (proteins) involved in DNA replication (Table). The most extensively studied examples include DNA replication from (i) *E. coli* chromosomal origin *ori* λ ; (ii) phage replication origin, *ori* λ (iii) phage P1 lysogenic origin *oriPl;* and (iv) ColEl and pBR322 plasmid origin. Among, these four systems, more variation is found in the initiation than in the elongation and termination of DNA replication.

Experimental Approaches for the Study of DNA Replication: Three different approaches have been used for the study of DNA replication in prokaryotes : (i) Isolation of mutants. Replication defective conditional mutants have been selected, which are, temperature sensitive dna_{ts} (non-conditional mutants, inhibiting DNA synthesis are usually lethal and are therefore not useful for a study of DNA replication). DNA synthesis in these mutants stops, when shifted to non-permissive temperature (e.g. 45°C). These can be either the replication initiation/termination mutants (also called 'slow stop mutants') which do not stop DNA synthesis immediately after shifting to non-permissive temperature (e.g. *dnaA, dnaB* and *dnaC),* or may be replication extension/elongation mutants (also called 'quick stop mutants'), which stop DNA synthesis immediately after shifting (e.g. *dnaB, dnaC, dnaE, dnaG, dnaN* and *dnaX).* Thus the proteins synthesized by the former class must be involved in initiation/termination and those synthesized by the latter must be involved in elongation.

In former class, DNA synthesis does not stop immediately, because DNA synthesis continues for one complete round on replication units where DNA synthesis was already initiated. Some of the viable mutants, which are not temperature sensitive, but are defective in DNA replication, have also been found useful. (ii) Use of inhibitors. Sensitivity of the initiation of DNA synthesis to rifampicin suggested that RNA synthesis is involved in the initiation of DNA replication. 'Novobiocin' and 'nalidixic acid' inhibited elongation by inhibiting 'DNA gyrase', which functions as a replication swivel. (iii) Isolation of proteins. Several proteins, which have a role in DNA synthesis, have been identified, by either their ability to complement the defective mutants or by their requirement for *in vitro* replication of plasmid or phage DNA. Genes for several of these proteins have now been isolated and cloned.

Initiation of DNA Replication

Initiation of Replication in Double Stranded DNA : A comparison of the initiation pathways used at *oriC, ori*λ, *oriPl* and *ColEl* (studied *in vitro)* suggests three steps : (i) recognition of the origin; (ii) opening of duplex to generate a region of single stranded DNA and (iii) capture of DnaB protein (a helicase). The events for initiation are sometimes also classified into 'prepriming' (which occurs only at the origin), 'priming' (which recurs with the initiation of each Okazaki fragment during elongation phase) and the formation of pre-initiation and initiation complexes. There are variations in different cases, in the mechanism of initiation, and also in the proteins involved in initiation. Only *E. coli oriC* will be described here, and differences will be indicated, wherever necessary and possible.

Prepriming and Priming : For initiation from *oriC*, following steps are involved (as deduced from *in vitro* studies) : (i) *DnaA-ATP complex* (10-20 monomers of DnaA) binds at 9bp inverted repeat regions (R_1, R_2, R_3, R_4) of *oriC* and promotes opening of the DNA in a region of three direct repeats of a 13-bp sequence (called *13-mers;* see Fig.). The opening occurs from right 13-*mer* leftwards, and requires negatively supercoiled DNA and Hu or IHE proteins, besides DnaA-ATP complex, (ii) DnaB is transferred to exposed single stranded DNA, with the help of DnaC, in presence of ATP. DnaC, perhaps, recognizes DnaA and thus helps in the transfer of DnaB, but DnaC itself is released from DnaB during the transfer reaction. (iii) 'DnaB helicase' causes unwinding of the DNA, on addition of ATP, SSB (single strand binding protein) and DNA gyrase. This leads to the formation of a prepriming complex. Unwinding and replication from *oriC* proceeds in both directions (bidirectional), SSB binding occurs on single stranded regions, and two DnaB complexes, are loaded one on each strand. (iv) Unwinding is followed by the synthesis of RNA primers by DnaG primase (through its interaction with DnaB) and the primers are elongated by 'DNA Pol III HE' (HE = holoenzyme). However, *in vivo,* priming of replication, in *E. coli* DNA, λ DNA and in the leading DNA strand of ColEl, is dependent on synthesis of RNA primer by RNA polymerase. (The mobile complex of helicase and primase has been termed a 'primosome' and the mobile complex of helicase, primase and polymerase at the replication fork, bringing about DNA replication is termed a 'replisome'). On both, *oriC* and *oriPl,* DnaA and DnaC proteins help in the transfer of DnaB. However in λ, phage, its own specific proteins λ *O* and λ P are available, along with many

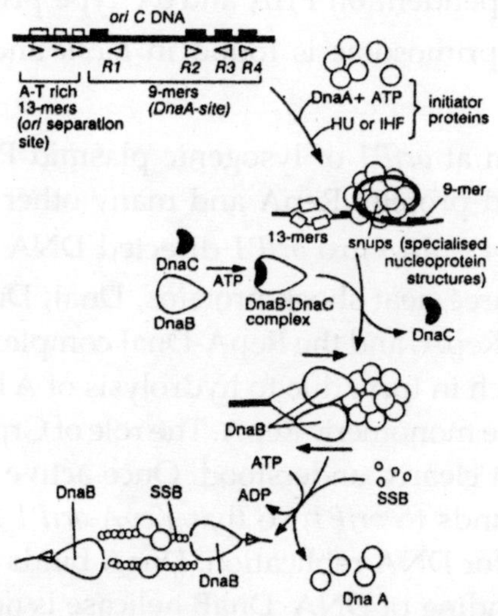

Model of the initiation of DNA replication at oriC of E. coli.

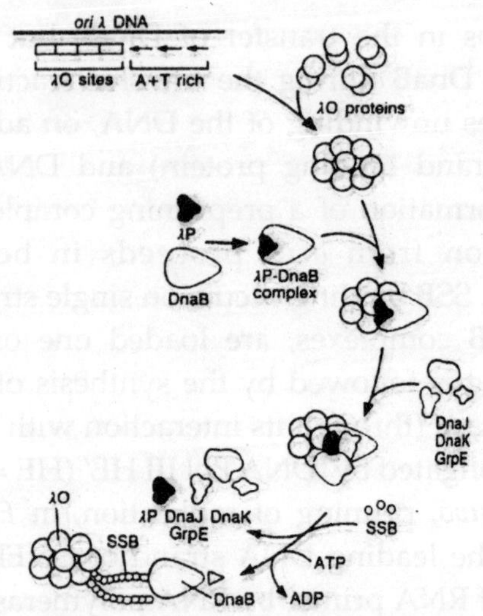

Model of the initiation of DNA replication at ari λ of phage λ.

E. coli proteins, for replication. *λO* and *λP* are analogous to *E. coli's* DnaA and DnaC respectively. Activation of DnaB in λ phage requires release of λ P, which is assisted by three heat-shock proteins, DnaK, DnaJ and GrpE (Fig.). Loading of DnaB on ColEl is dependent on PriA and ϕX type primosome. (The other type of primosome is found in *E.coli* and is known as *OriC* type).

Replication at *oriP1* of lysogenic plasmid P1 requires the phage-encoded protein, RepA and many other proteins used by *oriC* and *ori λ*. In vitro *oriP1* directed DNA synthesis also depends on three heat shock proteins, DnaJ, Dnak and GrpE. DnaJ binds to RepA, and the RepA-DnaJ complex is recognized by DnaK, which in turn, due to hydrolysis of ATP, triggers the release of active monomeric RepA. The role of GrpE in activation of RepA is not clearly understood. Once active in monomeric form, RepA binds to *oriP1,* so that *KepA-oriP1* DNA complex is a substrate for DNA replication. DnaA binds to its sites and leads to unwinding of DNA. DnaB helicase is now transferred to the origin, (perhaps through DnaC-DnaA interaction), and activation of DnaB helicase (without the involvement of DnaJ, Dnak and GrpE) leads to unwinding of the DNA. Now primase synthesizes primers, and DNA pol III HE catalyzes DNA synthesis (Fig.). *In vitro*, the replication at *oriPl* is unidirectional.

Initiation of DNA replication from ColEl origin is catalyzed by *E. coli* proteins, but differs in several essential features. On the leading strand RNA polymerase synthesizes the transcript forming RNA-DNA hybrid, that is processed by RNase H generating 3' OH group, which serves as a primer for elongation by DNA polymerase I forming DNA-DNA hybrid and displacing the other strand of the plasmid DNA. The displaced strand exposes a site known as 'primosome assembly site' or *pas*. A protein-DNA complex, called 'ϕX-type primosome' (including DnaB) is assembled at this site and is recognized by DnaG primase for initiation of primer synthesis. Thus the primer synthesis in ColEl differs on two strands and the replication is unidirectional. Eventually DNA pol III HE replaces DNA pol I on the leading strand and DnaG primase on the lagging strand for DNA elongation (Fig.).

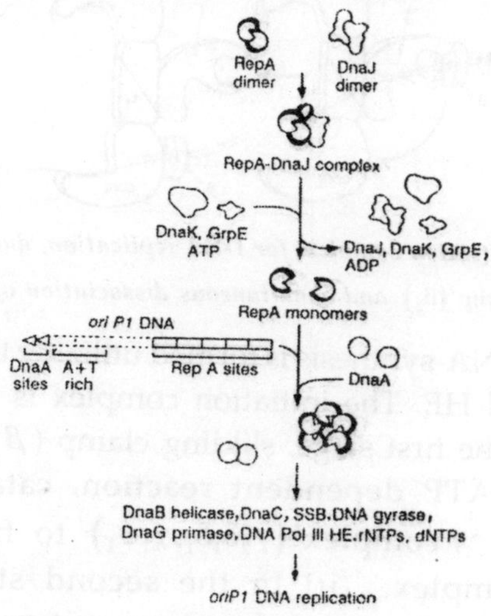

Model of the initiation of DNA replication of oriP1 of phage P1.

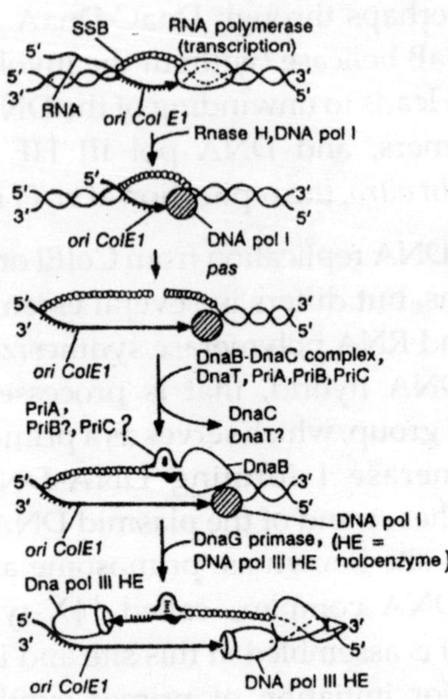

Model of the initiation of DNA replication at ColE1/pBR322 origin.

Formation of Pre-initiation and Initiation Complexes : As mentioned earlier, after prepriming and priming, an initiation

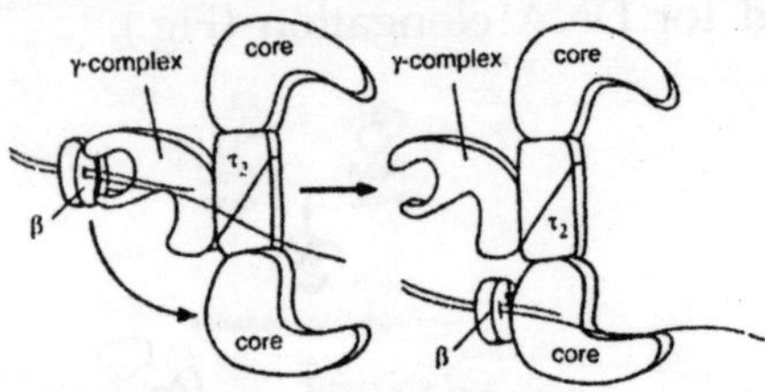

Formation of 'initiation complex' for DNA replication, due to repositioning of 'sliding clamp'(β_2) and simultaneous dissociation of 'γ complex'.

complex for DNA-synthesis is formed utilizing the components of DNA polIII HE. The initiation complex is formed in two stages: (i) In the first stage, slilding clamp (β_2) is loaded on DNA in an ATP dependent reaction, catalyzed by the matchmaker 'γ complex'($\gamma_2\delta_1\delta_1,\chi_1\psi_1$) to from the 'pre-initiation complex.' (ii) In the second stage, the core enzyme ($\alpha\varepsilon\theta$) is assembled with β_2 bound DNA to form the 'initiation complex' with simultaneous dissociation of 'γ complex' from the sliding clamp (β_2) (Fig.). The 'γ complex' though released from the sliding clamp (β_2), still remains associated with core $(\alpha\varepsilon\theta)$ and (β_2) on DNA through τ_2. In the holoenzyme, τ_2 acts as a bridge between 'core' and 'γ complex' to hold them together. The 'sliding clamp' (β_2) with double stranded DNA is now repositioned in contact with core, but away from 'γ complex' (Fig.). The retention of 'γ complex' in association with the rest of the enzyme, DNA pol III HE (at the replication fork) helps in multiple initiation events on the lagging strand. The protein and enzyme complex associated with the initiation complex formed at the replication fork, is also described as 'replisome'.

Initiation of replication in single stranded DNA phages. In many cases (including phage M13, which is single stranded and mtDNA) it is the RNA polymerase, which is involved in priming the DNA synthesis through the synthesis of small RNA segment, which precedes DNA synthesis. However, in phage G4 and in phage $\phi \times 174$ (both single stranded DNA), in place of RNA polymerase, a much smaller protein called primase (coded by *dnaG* gene) is utilized. This primase enzyme synthesizes RNA from one unique specific site each in phages G4 and Ml3, but at multiple sites in $\phi \times 174$.

In $\phi \times$ l74 (as also in G4 and M13), single stranded DNA is first coated with SSB protein, which is converted to a prepriming intermediate by six proteins named PriA, PriB, PriC, DnaT, DnaB and DnaC. The prepriming complex of six proteins assembles at a point described as 'primosome assembly site' or 'pas' (hairpin formed due to intrastrand pairing), and dnaG primase can now synthesize short RNA molecules at several sites. The complex of proteins including primase at the priming site is called primosome. Among the prepriming proteins PriA seems to be responsible for selecting the site (pas) for the prepriming complex and for displacing SSB. The other proteins (PriB, PriC) have an unknown prepriming function. It has, however, been shown that in the absence of SSB coating, primase in association with DnaB protein can synthesize RNA primer without the need of other five proteins (PriA, PriB, PriC, DnaT and DnaC), but a prior coating of SSB would need these five proteins for the synthesis of RNA primer.

Elongation of DNA Chain—replication Elongation Proteins : As discussed above, initiation of DNA replication involves loading and activation of DnaB helicase. DnaB directs other necessary enzymes (particularly DnaG primase and DNA pol III HE) to the template and help in DNA elongation.

DnaB Protein. DnaB is the major helicase, although other helicases have been discovered. It occurs as hexamer and unwinds the template in 5'-3' direction during *oriC* replication *in vitro,* at a rate of 500-1000bp/sec. The direction of unwinding requires DnaB to be placed on lagging strand, ahead of the polymerase on the leading strand. It also activates primer synthesis by DnaG primase. Since a DNA helicase not only unwinds DNA, but also translocates along the DNA during replication, it is believed to be a 'motor protein', with which it has many similarities.

DnaG Primase. The primase (DnaG) synthesizes a unique oligonucleotide primer on SSB coated single stranded phage G4 DNA. The sequence recognized by primase is over l00nt (nt = nucleotides) long containing a secondary structure. Since primer synthesis requires activation by DnaB, the latter is often described as a 'mobile promoter' of primer synthesis.

DNA Polymerase III Holoenzyme : DNA pol III HE is the large multiprotein complex responsible for elongation of DNA replication in *E. coli.* It has at least 10 subunits, each with specific function (Table). The holoenzyme consists of a 'core polymerase' and its 'accessory factors', which are analogous to the components of phage T4 and eukaryotic enzymes. In the holoenzyme, β subunit is twice as abundant as the other subunits, and in the cell its abundance is 300 times that of the polymerase assembly. It has a 'sliding clamp' mode of DNA binding. It has similarity with T4 *gene 45* protein and human PCNA.

Comparison of Pol III holoenzyme with Other Replicases : DNA polymerase III holoenzyme of *E. coli,* as shown above, consists of the following three components (with 10 subunits) : (i) a polymerase (core), (ii) a sliding clamp (β_2) and (iii) a clamp loader (γ complex). These three components can be compared with corresponding components in replicases of eukaryotes (Table).

A Comparison of the Components of Replicases in* E. coli, *Eukaryotes and T4 Phage.

Components of replicases	*Organisms*		
	E. coli	*Eukaryotes*	*T4 phage*
Polymerase/ exonuclease	Core (3 subunits)	Pol δ (2 subunits)	g43p (1 subunit)
Clamp loader (matchmaker)	γ complex (5 subunits)	RF-C , γ complex (5 subunits)	g44/62p complex (2 subunits)
Sliding clamp	β_2	PCNA	g45p
Proteins interacting with clamp	Pol III, Pol II	Pol δ , & Pol. ε , others.	g43p (pol), RNA polymerase

Other Proteins for DNA Elongation : Other proteins involved in DNA elongation are (i) DNA gyrase (a topoisomerase), which removes the topological strain associated with DNA unwinding; (ii) SSB (single stranded DNA binding) protein, that binds tightly to single stranded DNA thus stabilizing the unwound DNA; (iii) RNase H, which removes RNA primers; (iv) DNA pol I, which is used for filling the gap created due to removal of RNA primers and (v) DNA ligase, which converts Okazaki fragments into a continuous strand.

***Replication Fork Movement* :** During initiation to elongation transition, DnaB (while travelling in 5'-3' direction) generates a replication fork by opening the DNA duplex (Fig.). The strand having DnaB will thus be the lagging strand template. DNA primase associates with DnaB and synthesizes multiple primers for lagging strand and single primer for leading strand. The dimeric DNA pol III HE interacts simultaneously with both template strands. The leading strand polymerase moves in the same direction as DnaB, while copying the opposite strand. For the synthesis of lagging strand, the DNA polymerase will have to work on the same strand to which DnaB is bound, but it travels in opposite direction. The helicase and primase activities may be provided by DnaB and primase or alternatively by ϕX-type primosome.

DnaB, primase and 'DNA pol III HE' function together in elongation, although DnaB and DNA pol III HE seem to be tightly bound to the fork and stay bound throughout the reaction, an attribute described by saying that DnaB and DNA polymerase assembly is processive. In contrast, primase, 'β_2' and 'γ complex' act distributively (they dissociate and reassociate i.e. do not stay bound throughout the reaction).

***Continuous Synthesis on Leading Strand* :** The leading strand is primed only once on each of the two parental strands in bidirectional replication. In bidirectional replication, the first primer synthesized can serve as the 'leading strand primer' and subsequent primers serve as 'lagging strand primers'. No special mechanism is, therefore, needed for priming of the leading strand in bidirectional replication, and the DNA polymerase complex on leading strand is extremely processive.

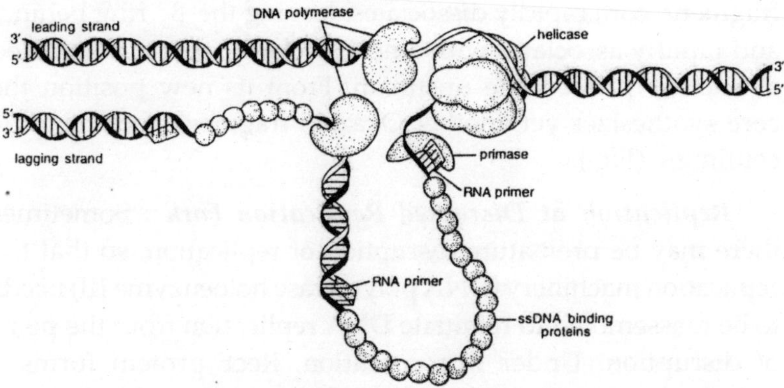

Three stages (a, b, c) in the synthesis of DNA on two strands and associated enlargement of loop due to the synthesis of leading strand.

However, in unidirectional replication an alternative mechanism for priming is needed for leading strand. For ColEl-type plasmids, RNA polymerase performs this role.

Synthesis of leading strand creates a loop of single stranded DNA, which is used as a template for the synthesis of lagging strand. This loop becomes larger, as the unwinding advances (Fig.). The length of this loop should be at least the size of an Okazaki fragment, before it can be used by the core enzyme for DNA synthesis.

Discontinuous Synthesis on Lagging Strand : The events on lagging strand are more complex than on leading strand, and DNA polymerase enzyme is less processive. Primase is taken up from solution and activated by DnaB, to synthesize a primer (10-20nt in length) on the lagging strand. The 'DNA pol III HE' interacts with primase and provides the signal for primase to dissociate from the fork. The primers are recognized by the polymerase on the lagging strand and are utilized for synthesis of 2000-4000 Okazaki fragments (each l-2kb in length) at a speed of lkb/second. This requires rapid recycling of 'core' polymerase, so that it finishes synthesis of an Okazaki fragment in one or two seconds and is recycled to the next RNA primer upstream. After completing the synthesis of an Okazaki fragment, core rapidly dissociates, leaving the β_2 ring behind, and rapidly associates with a new β_2 clamp, that is assembled at another primed site upstream. From its new position the core synthesizes yet another Okazaki fragment and the cycle continues (Fig.).

Replication at Disrupted Replication Fork : Sometimes there may be premature disruption of replication, so that the replication machinery (DNA polymerase holoenzyme III) needs to be reassembled to reinitiate DNA replication from the point of disruption. Under this condition, RecF protein forms a complex with RecO and RecR (RecFOR), which would recognize the disrupted replication fork. RecFOR facilitates the reassembly of a replication holoenzyme at this site, much the same way, as the DnaA protein recognizes the structure at *oriC* (Fig.).

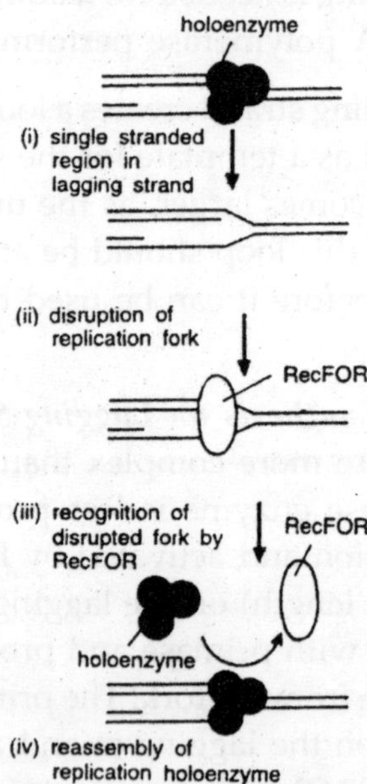

DNA replication at disrupted replication fork, due to reassembly of holoenzyme with the help of RecFOR.

Termination of DNA Replication : The *E. coli* chromosome carries a large termination zone, diametrically opposite from *oriC,* which blocks the progress of replication forks (of bidirectional replication), meeting at this region. *E. coli* chromosome and several plasmids carry specific sequences, called *ter* sites, where TBP (ter binding protein) or 'Tus protein' binds. [In *B. subtilis,* the protein equivalent to Tus is called replication terminator protein (RTP)]. In the termination zone of *E. coli,* there are three ter sites *(terA, terD* and *terE)* for counter-clockwise fork and three *ter* sites *(terB, terC* and *terF)* for clockwise fork. These six sites are arranged in overlapping manner, leaving no 'replication-free' gap on the chromosome (Fig.). TUS-ter complexes formed at *'ter'* sites stalls the replication fork, by inhibiting the DNA helicase or DnaB. When this termination zone is deleted, replication stops simply by the meeting of opposite replication forks, suggesting that the termination zone is not essential.

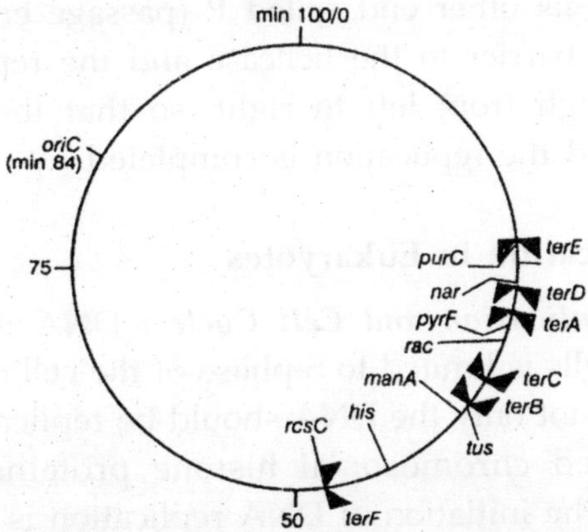

A partial map of E. coli chromosome, showing the position of six ermination sites (terA-terF) and the tus gene relative to the location of oriC.

In 1996, X-ray crystal structure of (Tus)-ter complex from *E. coli* was published. The protein-DNA interaction was found to be completely new and explained how the complex arrested the replication fork on its one end (B or blocking end) and not on its other end (P or passage end). The Tus protein in the (Tus)-ter complex is organized into two domains, connected by two antiparallel beta strands, thus creating a central cleft into which DNA fits (Fig.). The Tus protein grasps DNA, and on the right hand provides a barrier (B end of the complex) to the replication fork moving from right to the left. This means that the helicase can not disrupt the tight *Tus-ter* contacts. The

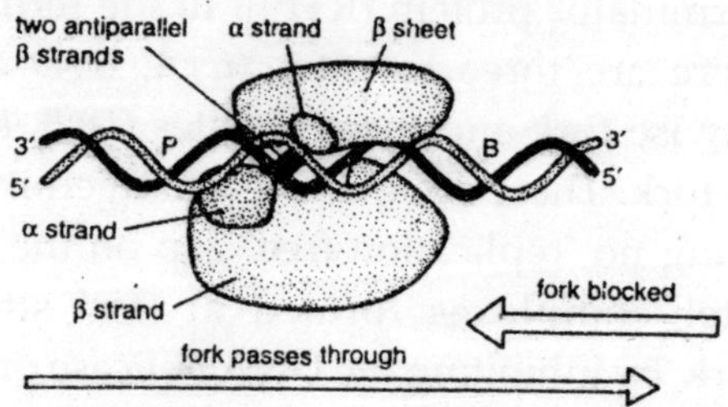

The interaction of Tus protein with ter sites as revealed by the three dimensional structure of (Tus)—ter complex (see text for details) for termination of DNA replication.

complex, on its other end called P (passage end), does not provide any barrier to the helicase and the replication fork moves through from left to right, so that the complex is displaced and the replication is completed.

DNA Replication in Eukaryotes

DNA Replication and Cell Cycle : DNA replication in eukaryotic cells is limited to S-phase of the cell cycle. During the S-phase, not only the DNA should be replicated, but also the associated chromosomal histone proteins should be duplicated. The initiation of DNA replication is a prominent landmark in the cell cycle and is, therefore, an important point for exercising control. Further, while in *E. coli* and other prokaryotes, a solitary origin for DNA replication is found, in eukaryotes, DNA replication needs to be initiated at multiple origins (up to several thousands; 10^3 to 10^5 replication events occur in each cell cycle) in a coordinated manner, though not identically at all origins. This leads to great variation in the duration of S-phase (S phase is 3.4 minutes in early embryonic cells, but 10 hours in cultured cells in *Drosophila melanogaster)*, even though the rate of DNA synthesis does not differ. It is possible that different specific initiator proteins are used at different origins at different times during S-phase. For instance, in yeast, a protein Abflp (a transcription factor) binds many origins, but not all origins of replication. However, a set of origins may be organized in clusters to be used synchronously.

Replication Origins and Initiaton of DNA Replication (cis and Transacting Elements: Repliators and Initiators) : For a study of replication origins, eukaryotes have been classified into (i) simple genomes including those from yeast and a number of eukaryotic viruses and (ii) complex genomes, represented by specific regions of chromosomes belonging to *Xenopus*, Chinese Hamster Ovary (CHO) cell lines, *Drosophila* and other metazoans. While more detailed information is

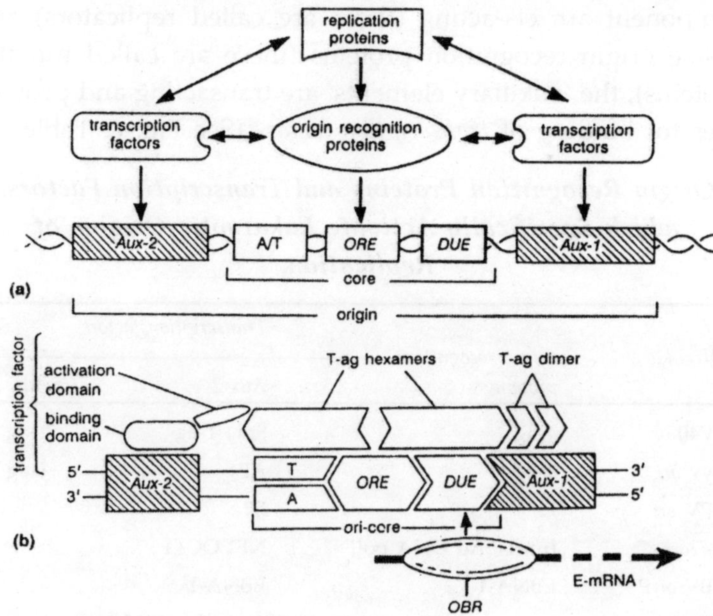

Details of a eukaryotic origin of DNA replication : (a) different components of ori; (b) binding of proteins on different components of ori (for details, see text)

available for simple genomes, the information for complex genomes is both fragmentary and difficult to explain. In simple genomes, an origin of replication encompasses 100-l000bp of DNA (100-200bp in budding yeast; 500-l000bp in fission yeast), but in complex genomes DNA replication occurs at many sites within an *initiation zone*, although some sites are highly preferred over others.

In simple genomes, *ori* consists of two parts : (i) core component (analogous to transcription promoter); it consists of an *Origin Recognition Element (ORE)*, a *DNA-Unwinding Element (DUE)* and an *A/T-rich element*, (one strand is A-rich and other strand is T-rich; the A/T rich element is part of ORE in some cases and it serves as DUE itself in others); (ii) one or more auxiliary components e.g. *Aux-1* and *Aux-2* (analogous to transcription enhancers). While the sequences of the core component are cis-acting (these are called replicators) and utilize origin recognition proteins (these are called initiator proteins), the 'auxiliary elements' are transacting and provide sites for binding of *transcription factors* (See Fig. & Table).

Origin Recognition Proteins and Transcription Factors, which Specifically Activate Eukaryotic Origins of Replication.

Origin	*origin recognition protein(s)*	*Transcription factors*	
		Aux-2	*Aux-1*
SV40 *ori*	T-ag	Spl., T-ag	T-ag
Pyv *ori*	T-ag	AP1	T-ag
BPV *ori*	El	E2	
Ad2ori	[preTP:Ad DNA pol]	NF1 OCT1	
EBV *oriP*	EBNA-1	EBNA-1	
HSV *oriH*	UL9	Several candidates	
mtDNA *oriH*	RNase MRP	mtRNA polymerase	
Yeast ARS	ACBP?, ORC?	ABF1	ABF1

Among simple eukaryotic genomes, replication origins in yeast *(Saccharomyces cerevisiae)* were studied in great detail and were termed 'replicators' or 'autonomously replicating sequences (ARSs)'. It has been shown that these sequences, when present in a plasmid, help high frequency of transformation and allow autonomous replication of plasmids in yeast cells. It has been shown that all DNA replication origins in yeast cells are ARSs, although all ARSs are not replication origins. (In all ARSs, meant for DNA replication,

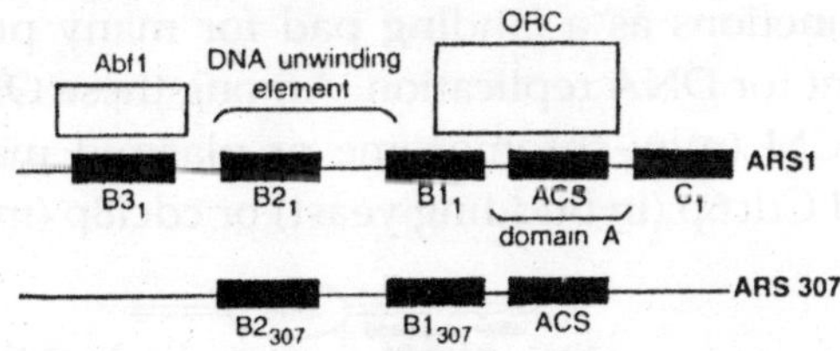

Components of two different auto Inomously replicating sequence elements (ARS1, ARS 307), as DNA replication origins in yeast.

there is an 11bp *'AT-rich consensus sequence'* or *'ACS'* or *'domain A'*, which is the core component, and is essential but not sufficient for the function of ARS. The two ARS elements that have been studied in detail are ARS1 and ARS307. Both have domains *A, B* and C. The domain *B* is divided into subdomains *Bl, B2* and *B3* in ARS1 and into subdomains *B1* and *B2* in ARS307 (Fig.). Although there is no sequence similarity, but there is functional conservation between these two ARSs (outside ACS). At 3' end of the T-rich strand of ARS, the *'domain B'* is believed to be involved in binding of initiator protein, helix destabilization via DNA unwinding, DNA bending, and binding of accessory protein factors.

The sequences 5' to the T-rich strand of the ACS are similarly referred to as *'domain C'*. The domain A of ARS functions in concert with subdomains of domain B, each of which is important but not essential for origin's activity. The transcription factor Abflp (ARS binding factorl) binds at B_3 in

ARS1 (analogous to auxiliary components) and trans-activates ARS elements, up to lkb away from 'ACS' or 'domain A'.

ORC, MCM, Cdc6p/Cdc18p and Licensing Concept : A search for initiator factors that recognize ARS1 (replicator) in a sequence-specific manner, revealed that a complex of six polypeptides binds ACS element (A) of ARS1 and Bl element with high specificity. This complex is described as 'origin recognition complex (ORC)' and is a eukaryotic 'initiator'. There are also proteins which bind B_3 and B_2-. It is believed that ORC functions as a landing pad for many proteins that are important for DNA replication. Among these ORC binding proteins, MCM (mini-chromosome or plasmid maintenance) proteins and Cdc6p (in budding yeast) or cdcl8p (in the fission

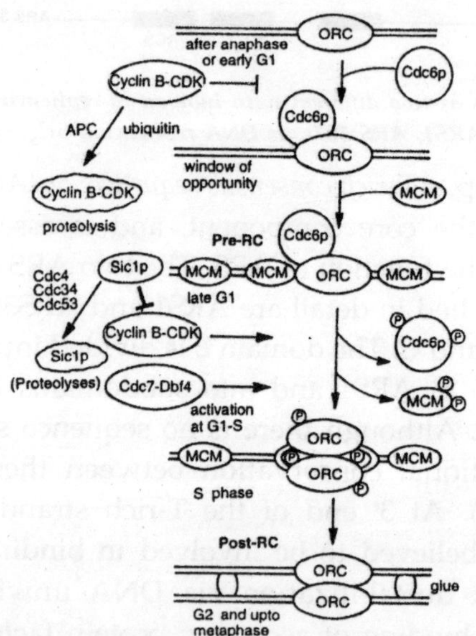

The role of ORC, MCM and Cdc6p/Cdc18p in granting licence for DNA replication in yeast (see text for details).

yeast) are the most important. These MCM proteins are essential for the maintenance of some 'ARS containing plasmids'. It is also speculated that perhaps Cdc6p/Cdcl8p is the loading factor (RLF-B; RLF = replication licensing factor) for the MCM complex at late G1. Further it is believed that MCM complex, in association with ORC at the origin site must be phosphorylated (by kinases, Cdc7p-Dbf4p, cyclin B-CDK), to initiate replication, thereby removing MCMs and thus granting licence for replication (Fig.). This is described as 'licensing concept'.

There are two states of the chromosome during DNA replication : (i) a pre-replication complex (pre-RC) state (with ORC, Cdc6p, MCM, etc.), corresponding to competent state of G1 chromosomes and (ii) a post-replication complex (post-RC) state containing only ORC. The pre-RC is activated by cell-cycle regulated activators to cause initiation of DNA replication and is disrupted after the initiation of DNA replication.

The 'origin recognition proteins' listed in Table, help directly in DNA unwinding and indirectly in the initiation of replication through their association with 'replication proteins' including DNA polymerases. The transcription factors, which bind the 'auxiliary elements', have each a 'DNA binding domain' and an 'activation domain', the latter involved in activation of 'origin recognition proteins' through protein-protein interaction (Fig.). The transcription factors thus facilitate DNA replication through the following activities : (i) initiation of transcription through *ori* core to produce RNA primers for DNA synthesis; (ii) the assembly of an initiation complex; (iii) activation of initiation complex and (iv) overcoming repression due to chromatin structure.

In complex genomes, a surprising feature is the appearance of 'replication bubbles' (100-500bp) throughout an 'initiation zone' (5000-55000bp). This suggests that replication starts at many sites within an initiation zone.'

There is, however, evidence that most replication events in metazoan chromosomes begin at sites, which are not randomly chosen, but are specific sites within the larger initiation zone, which is defined by the presence of replication bubbles. There is also evidence that DNA synthesis occurs by the replication fork mechanism and nucleosome segregation is distributive. The most extensive analysis of mammalian DNA replication (up to 1992) has been carried out in a 300kbp region around the *dhfr* gene of Chinese Hamster Ovary (CHO) cells (dhfr = dihydrofolate reductase). In this 300kbp region, replication begins in three different initiation zones referred to as *Ori-*α, *Ori-*β, *Ori-*γ. In each of these initiation zones, the origin of replication is described as an 'Origin of Bidirectional Replication (OBR)' associated with several elements, which facilitate replication initiation (Fig.). The *Ori-*β, which is the most extensively studied zone in this region is flanked by two 'matrix attachment regions *(MAR)*'. The *OBR* within *ori-*β is flanked by (i) two *Alu* repeats, (ii) sequences associated with DNA amplification, (iii) autonomously replicating plasmid

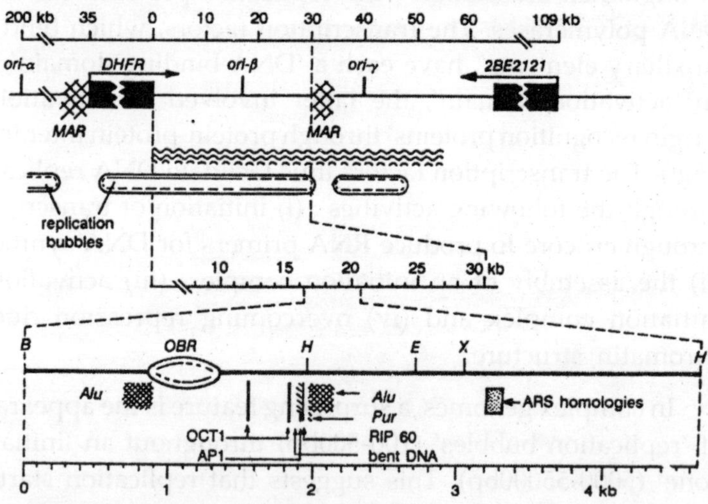

*Details of dhfr region (in a CHO cell line) with three initiation zones (ori-*α*, ori-*β*, ori-*γ*); lower figure shows details of ori-p (for details, see).*

DNA, (iv) a segment of bent DNA, (v) sequences homologous to yeast ARS consensus sequence, (vi) binding sites for at least four proteins involved in the initiation of DNA replication

(RIP60, PUR, 0CT1 and AP1; the last two are transcription factors).

Although, the mechanism of DNA replication in complex genomes like those of mammals is not fully understood, some generalizations can be made: (i) initiation of DNA in mammalian cells generally occurs in intragenic regions and initiation zones are much longer as indicated above; (ii) the zone where replication is initiated changes during development, this being broad in early embryonic and restricted in later stages (as shown for ribosomal DNA in *Xenopus);* (iii) there are also DNA sequences, called locus controlling region (LCR), which can control positioning of origins of DNA replication to specific sites (as shown for human β-globin gene).

CDKs and Cyclins in DNA Replication : Cyclin-CDK activity is also needed both for initiation of DNA replication at START point of cell cycle and also for blocking the initiation of DNA replication during and after the S phase of cell cycle. For instance, in budding yeast, B-type/S-phase cyclins (C1b5p, C1b6p) are needed for DNA replication. The activity of C1b5-CDK (e.g. C1b5/6-Cdc28) is inhibited by Sic1p, which, therefore, needs to be removed to allow initiation of DNA replication. The removal of Sic1p is achieved through ubiquitin-dependent proteolysis after START. This proteolysis requires Cdc34p, Cdc53p, Cdc4p and Skp1p proteins and also the G1 cyclin-CDKs (C1n1p-Cdc28p, C1n2p-Cdc28p). Thus cyclins and CDKs have been shown to have multiple roles, particularly including the following: (i) activate DNA replication; (ii) prevent re-replication by blocking the formation of pre-RC in G2 and (iii) allow establishment of metaphase. After mitosis, the cycle is reset, by degradation of cyclins due to anaphase promoting complex (APC) and ubiquitin dependent proteolysis.

Cdc7p-Dbf4p for MCM Phosphorylation : The DNA replication in eukaryotes requires another kinase activity (independent of cyclin-CDK) made available through Cdc7p-

Dbf4p, which activates competent G1 state. The Cdc7p protein (conserved in yeast to mammals) is a kinase, whose activity depends on Dbf4p (it also interacts with Cdc5p, another kinase) and remains active during G1 to S phase. It is believed that Cdc7p phosphorylates MCM proteins in the pre-RC.

Cln3-Cdc28 Kinase for Transcriptional Control of S Phase genes: Many genes (~30 genes) whose products are needed during S phase, are transcribed during late G1 phase. This late G1-transcription is regulated by C1n3-Cdc28 kinase. The transcription factor MBF, which binds to the MCB elements and the transcription factor SBF, which binds the SCB promoter element, are activated by C1n3-Cdc28 kinase to stimulate expression of some important S-phase genes including B-cyclin genes *(Clb5, Clb6)* and Gl-cyclin genes *(Clnl,Cln2)*. Thus Cdc7p-Dbf4 and Cln3-Cdc28 kinases trigger the initiation of DNA replication.

Comparison of Initiation of DNA Replication with Transcription Initiation : The initiation of DNA replication has also been compared with transcription, due to many similarities: (i) both involve ordered protein-DNA and protein-protein interactions leading to site identification and positioning of appropriate polymerase at this site; (ii) transcriptional promoters are heterogeneous regulatory elements and so are replication origins; (iii) promoters have common core elements, where basic transcription factors bind; replication origins also have core consensus sequences for binding of recognition factors like ORC; (iv) promoters have flanking sequences besides core elements; accessory factors binding on these flanking sequences, modulate the activity of basic transcription factors and thereby regulate the rate of transcription initiation in response to a variety of stimuli; at replication origins also there are 'accessory control elements' (e.g. B_3 at ARS1) or 'auxiliary elements', where transacting factors bind and modulate the activity of the replication origin. Since transcription factors are involved both in transcription and replication, they may play an important role in temporal coordination of transcription and replication during the cell cycle (Fig.).

Different Steps Involved in Eukaryotic DNA Replication : During 1987-1989, on the basis of studies on classical enzymology, a proposal was put forward for eukaryotic DNA

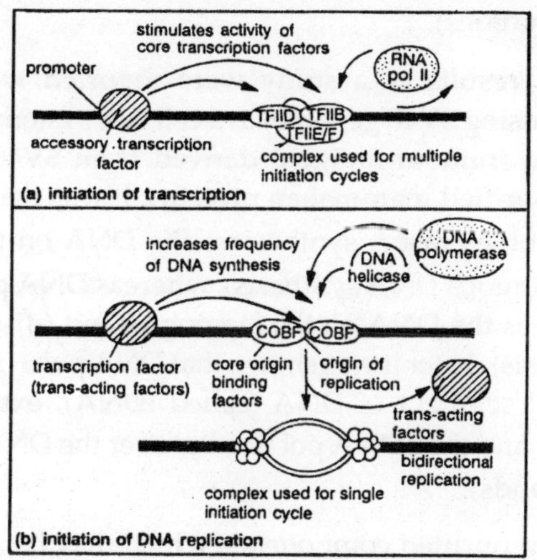

A comparison of initiation of DNA replication with transcription initiation in eukaryotes

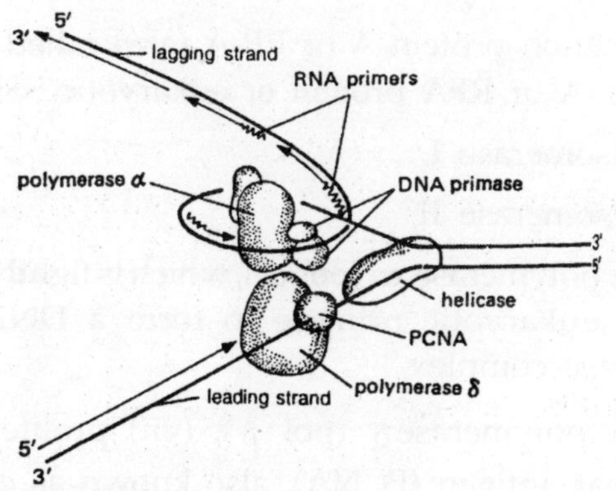

Two different DNA polymerases, synthesizing leading and lagging strands at a eukaryotic DNA replication fork

replication. According to this proposal, unlike DNA replication in prokaryotes, two different DNA polymerases were required for eukaryotic DNA replication (Fig.). These two DNA polymerases (DNA polymerase α and DNA polymerase δ) were purified and characterized. (Corresponding enzymes in yeast are called DNA polymerase I and DNA polymerase III, which should not be confused with *E.coli* enzymes bearing same designations).

In 1990, results of a study were reported where DNA replication (using SV40 genome) in a cell free system depended on T antigen (tumour antigen) derived from SV40 and on a number of purified mammalian proteins. It was demonstrated that DNA polymerase δ synthesizes the DNA on the leading strand (continuous DNA synthesis), whereas DNA polymerase α synthesizes the DNA on the lagging strand (discontinuous DNA synthesis). Later it was shown that DNA pol α synthesizes only a small segment of DNA (called iDNA), extending the RNA primer and then DNA pol δ takes over the DNA synthesis on both strands).

The eight purified components, which may be involved in DNA replication, included the following:

(i) T antigen,

(ii) replication protein A or RP-A (also called replication factor A or RFA protein or eukaryotic SSB),

(iii) topoisomerase I,

(iv) topoisomerase II,

(v) DNA polymerase α (pol α), which is tightly associated with eukaryotic primase to form a DNA poly α-primase complex,

(vi) DNA polymerase δ (pol δ), (vii) proliferating -cell nuclear antigen (PCNA), also known as cyclin,

(viii) replication factor C or RF-C (ATPase).

Of these eight components, T antigen (TAg) was the only component, which was SV40 encoded. Other components were derived from the host mammalian system. A comparison of different proteins used for DNA replication in *E.coli,* HeLa cells/SV40 and the phage T4 is presented in Table).

TAg has a DNA binding domain and a domain having ATPase and helicase activities. The SV40 *ori,* involved in the initiation of DNA replication, consists of three defined regions extending over 65 bp :

(i) an imperfectly inverted repeat sequence,

(ii) a 27 bp binding site for TAg (site II) and

(iii) a 17bp AT-rich sequence, meant for DNA unwinding.

Adjacent to 65 bp sequence, there is another TAg binding site called site I. These features may be shared by many or all eukaryotic DNA replication origins. Following steps are suggested for eukaryotic DNA replication, (i) Before DNA synthesis, there is a presynthesis stage of 8-10 minutes duration, for the formation of unwound DNA complex. This stage requires only three purified proteins, namely T antigen (TAg), RP-A and topoisomerase I or II.

The T-antigen, using its DNA binding domain, forms a multi-subunit complex with site I and site II in the presence of ATP and causes local unwinding. More extensive unwinding follows the association of RP-A and a topoisomerase, with the help of DNA helicase component of TAg. In a number of other viruses also, proteins comparable to TAg are used for the formation of unwound complex for initiation of DNA replication. The topoisomerases help in unwinding of DNA by altering topology of DNA at the replication fork. RP-A or SSB binds to unwound single stranded DNA. (ii) The primase, which is tightly associated with DNA polymerase α , undertakes primer RNA synthesis, (iii) The above priming

Different Proteins Required for Similar Functions at the Replication Forks in Bacterial and Mammalian Systems

	Function	Organisms			
		E. Coli	*Archaea*	*HeLa/SV40*	*Phage T4*
1.	Helicase	DnaB	Dna 2 like, MCM 2 like	T-antigen	g47p
2.	Loading belicase/primase	DnaC	?	T-antigen	g59p
3.	Single strand maintenance	SSB	RPA-klike	RPA	g32p
4.	Priming	DnaG	Pol α -like	Pol α /primase	g61p
5.	Sliding clamp	β dimer (β_2)	PCNA-like	PCNA	g45p
6.	Matchmaker or clamp loading (ATPase)	'γ complex'	RFC-like	RFC	g44/g62p
7.	Catalysis	Pol III core	Family B DNA	Pol δ	g43p
8.	Holoenzyme dimerization	τ dimer (T_2)	1	?	g43p
9.	RNA removal	Pol I	FEN1	MFI	g43p
10.	Ligator	Ligase	Ligase	Ligase T	T4 ligase

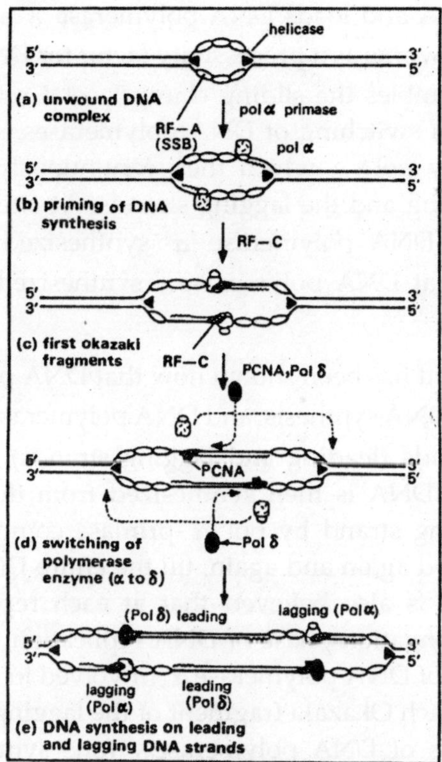

Model for initiating leading and lagging strand synthesis with two DNA polymerases. The first line of the figure shows the unwound complex stage and then four subsequent stages : primer RNA synthesis with primase (priming), first Okazaki fragment synthesis with pol α , switching of DNA polymerases at the 3' end of this first Okazaki fragment, and finally, leading and lagging strand synthesis by pol δ and pol α are indicated (see text for details). Factors involved are DNA helicase (T antigen in the case for SV40), RF-A, pol α -primase, RF-C, PCNA and pol δ . The factors are introduced according to the order of entry into the reaction. Wavy lines are, synthesized primer RNAs, and the dashed lines indicate the introduction of pol δ and the movement of pol α -primase complex along the lagging strand template without dissociation from the replication fork.

reaction is unusual; it starts with RNA synthesis, like other priming reactions, but the RNA primer is extended by the DNA polymerase activity of DNA pol α , to give *a* short DNA sequence (3-4 bases long), called iDNA. (iv) RF-C binds to the

3' end of iDNA and loads DNA polymerase δ and PCNA (a cyclin, which works as a processivity factor for DNA pol δ and therefore resembles the sliding clamp β_2 of *E.coli*). RF-C and PCNA help in switching of DNA polymerases, so that pol α is replaced by pol δ , which then continuously synthesizes both the leading and the lagging strands. At one time, it was thought that DNA polymerase α synthesized the lagging strand and that DNA polymerase δ synthesized the leading strand.

However, it has been shown now that DNA polymerase α only initiates DNA synthesis, and DNA polymerase δ elongates both the strands (leading and lagging strands), (v) Another RNA primer-iDNA is then synthesized from the replication fork on lagging strand by pol α -primase complex and this step is repeated again and again, till the entire DNA molecule is covered. It is also believed that at each replication fork during the elongation phase of DNA replication, there is only one molecule of DNA polymerase α (involved in the synthesis of iDNA for each Okazaki fragment of the lagging strand) and two molecules of DNA polymerase δ (one synthesizing the leading strand and the other synthesizing the lagging strand). (vi) The RNA primers are removed using MF1 (a (5' - 3') exonuclease) and the gaps are filled by DNA polymerase δ , as by DNA pol I in prokaryotic DNA replication. DNA ligase I seals the sticky ends after gap filling. This scheme of DNA replication involving switching between pol α and pol δ is believed to operate in all eukaryotes (Fig.).

In another study, role of yeast DNA polymerase ε in DNA replication was suggested, so that three DNA polymerases (α , δ and ε) were known to be involved in DNA replication. To accommodate three DNA polymerases functioning at the replication fork, A. Sugino and his colleagues proposed that polymerase α might function only as an initiation polymerase at both leading and lagging strands (only polymerase α has primase activity), whereas polymerase ε and polymerase δ are involved in elongation of the leading and lagging-strands respectively. There are other models, which suggest the role of either the polymerases α and δ or only polymerase α in eukaryotic DNA replication. It is also possible that in different systems, different enzyme systems are involved.

Synthesis of Telomeric DNA by Telomerase : This feature is common to all chromosomes of a species and is conserved. A special ribonucleoprotein molecule (enzymatic in nature) called telomerase uses a special mechanism for the synthesis of DNA at these telomeric ends.

The DNA repeat sequence of telomere has one G-rich strand and the other C rich strand, the G-rich strand having a single stranded overhang. This overhang works as a primer and for its elongation uses as template the RNA component of telomerase enzyme (Fig.). Telomerase synthesizes only the G-rich strand of telomeres. The complementary C-rich strand is perhaps synthesized (extended) by primase-polymerase mediated discontinuous synthesis, typical of semi- conservative DNA replication for which extended G-rich strand is used as a template. For the use of telomerase RNA as template, its sequence should be complementary to telomeric DNA repeat unit, which has been verified. For instance, in *Euplotes*, 5'CAAAACCC CAAAA3' is found in telomerase, which is used for synthesis of 5'GGGGTTTT3' repeats.

The protein component of telomerase has also been studied. It has reverse transcriptase activity to be able to synthesize DNA on RNA template. In 1997, the protein carrying this reverse transcriptase (RT) activity and the gene encoding this protein were identified in yeast. These proteins were designated p123 and Est2 and the gene(s) encoding these proteins were called *EST* (ever shorter telomeres) genes. These proteins having RT activity are also being searched in organisms other than yeast.

Regulation of Telomere Length : In recent years, telomeres have been shown to undergo alteration in their lengths under specific conditions. Following are some examples: (i) telomeres are shorter in tumours; (ii) telomeres shorten, when cells divide in culture, (iii) in certain human tissues, telomeres are shorter in older people than in younger people; (iv) telomeres are longer in sperm cells. Since telomere shortening is associated with senescence and cancer, effort is also being made to establish

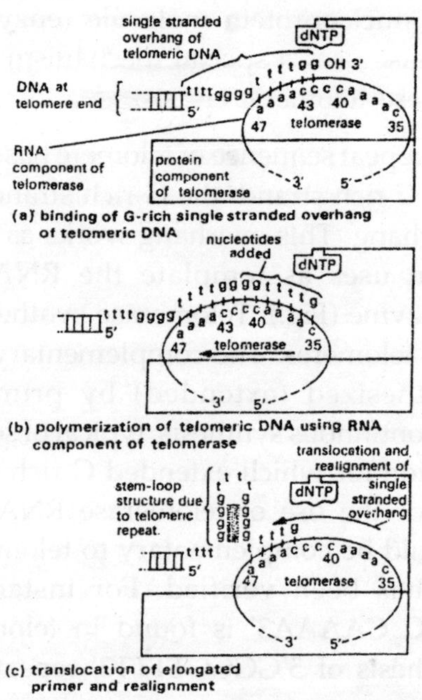

A model for the action of the telomerase from the ciliate Euplotes. This model is based on a model first proposed for the Tetrahymena telomerase, but in the Euplotes telomerase, the template nucleotides have been more accurately defined by in vitro studies, (a) the telomeric primer is bound by interactions with the telomerase through recognition of its G:C base-paired secondary structure and Watson-Crick DNA-RNA base pairing between the 3'-nucleotides of the primer with the telomerase RNA template. (b) polymerization copies out to position 35 on the template. (c) the elongated primer translocates backwards, so its new 3' end, becomes aligned with the template, again by base pairing as shown. The unpairing of the RNA-DNA helix in this step is aided by G:C pairing (indicated by shading) within the newly elongated primer. Another cycle of DNA polymerization occurs, further elongating the primer.

whether or not telomeres play a causal role in entry to senescence and in causing cancer. Evidence however is available suggesting that reduction in telomere length may not cause aging; instead it may be the consequence of aging.

In view of the above, the processes involved in regulating the telomere length have been studied in some detail. Telomerase enzyme specifically elongates telomeres and there are following other factors that influence telomere length: (i) telomere binding proteins; (ii) telomere capping proteins and (iii) DNA replication enzymes.

Models of DNA Replication

Replication Fork Model : The most extensively utilized mechanism of DNA replication involves formation of replication forks, moving in one direction in unidirectional replication and in both directions in bidirectional replication. This method can apply to both linear as well as circular DNA molecules. Replication of circular double helical DNA through formation of replication forks moving in both directions (bidirectional replication) is shown in Figure.

Rolling Circle Model of DNA Replication : An alternative mechanism has been suggested for the replication of circular DNA and this is called rolling circle mechanism, which explains multiplication of many viral DNAs and also explains DNA replication during bacterial mating. One of the evidences in support of this model is the observation of long chains of polynucleotides which could measure several times the contour length of circular DNA.

In this mode, due to nicking in one of the two strands, one strand becomes a polynucleotide with two ends, one end having 3'—OH group, the other having 5'p group. DNA synthesis starts by adding bases at 3'—OH end, displacing the 5' end, which rolls out as a free tail, and is then copied in 5' $\rightarrow$ 3' direction in small segments using fresh enzymes system. These tails thus become double stranded and give rise to a circular DNA later on. A possible mechanism of replication of circular DNA using rolling circle model is shown in Figure.

Mitochondrial DNA Replication and D-loops : In some animal cells (vertebrates), D-loops have been observed in replicating mitochondrial DNA (mtDNA). With the help of radioactive label, the fate of these D loops was studied. It could be shown that D loops represent early stages of DNA replication in mitochondria. The replication model based on D-loop is shown in Figure. The twist in supercoiled duplex is transiently relieved so that a new strand can be initiated at a fixed point, using one of the two strands. The other strand, which is not copied, balloons out as D-loop and can be seen in electron microscope at a later stage. The replication starts later on displaced single stranded D-loop. In this manner replication

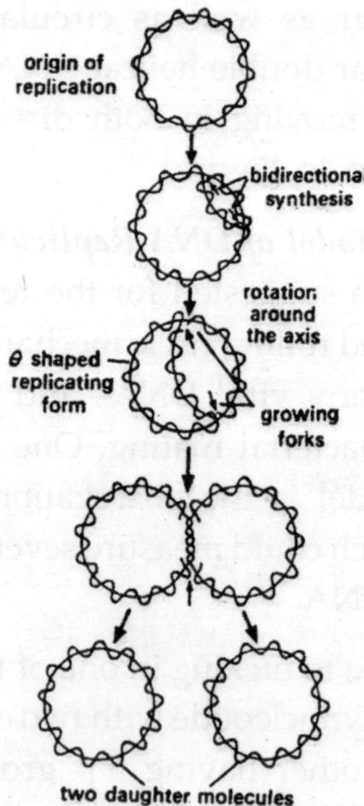

Model for replication of circular double helical DNA, based on formation of double forks due to bidirectional DNA replication.

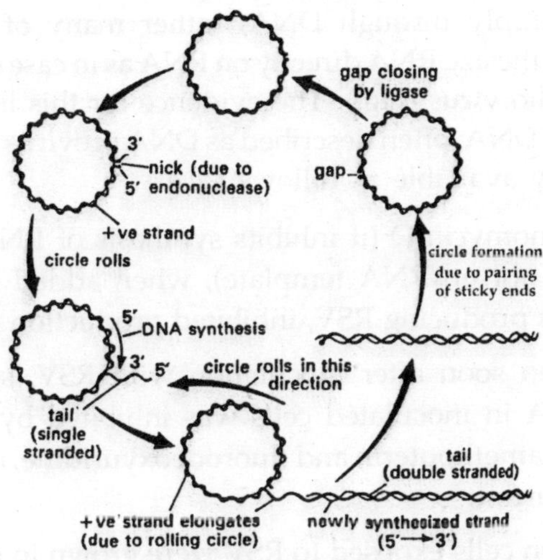

Model for replication of circular double helical DNA, based on rolling of DNA circle following a nick formation due to endonuclease.

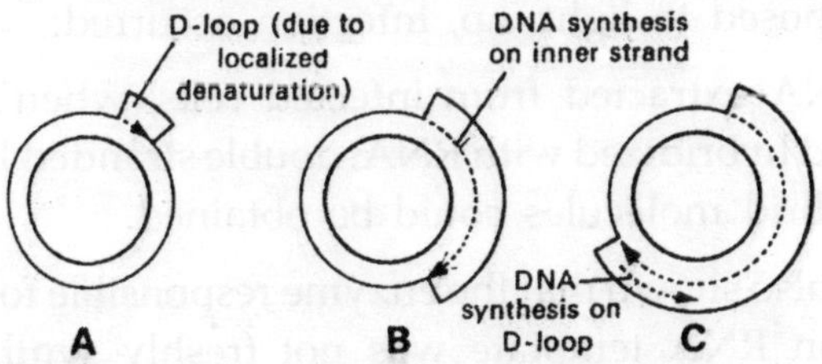

Model for replication of mitochondrial DNA based on D-loop; D-loop found at A, is extended in B, where replication takes place on inner strand. Replication on displaced strand forming D-loop starts later as shown in C.

is unidirectional until about two third of one strand (H-strand) is already synthesized. About 100 minutes are needed for completion of mtDNA replication, and DNA loops in various stages of replication can be seen at a given time.

RNA Directed DNA Synthesis (Reverse Transcription) : In early 1960's, it was shown that some RNA viruses (e.g. Rous Sarcome Virus = RSV), after infecting, the host, gives rise to DNA as an intermediate step in their multiplication. This DNA then gives rise to (i) RNA for assembly of viral genomes and (ii) mRNA for synthesis of virus specific proteins. Not all RNA viruses multiply through DNA; rather many of them may actually synthesize RNA directly on RNA as in case of influenza virus or polio virus (Fig.). The evidence for this intermediate synthesis of DNA, often described as DNA provirus hypothesis, was initially available as follows :

(i) actinomycin D (it inhibits synthesis of RNA on DNA but not on RNA template), when added to cultured cells producing RSV, inhibited production of all RNA;

(ii) when soon after inoculation with RSV, synthesis of DNA in inoculated cells was inhibited by antibiotics like amethopterin and fluorodeoxyuridine, no infection occurred;

(iii) when cells exposed to RSV were grown in presence of bromodeoxyuridine (its incorporation renders the newly synthesized DNA sensitive to light) and then exposed to light, no, infection occurred;

(iv) DNA extracted from infected cells, when denatured and hybridized with RNA, double stranded DNA-RNA hybrid molecules could be obtained.

It was also shown that the enzyme responsible for synthesis of DNA on RNA template was not freshly synthesized on infection, but was present within the virus particles (RSV).

The above information on RNA directed DNA synthesis was available in 1960's but no enzyme for this activity could be discovered. In 1970, for the first time, disrupted virus particles from RSV (used by H. Temin) and MLV (mouse leukaemia virus; used by D. Baltimore) were shown to undertake DNA synthesis, utilizing dNTPs (one of them made radioactive), suggesting that an enzyme existed in the core of virus particle, which could stimulate RNA-directed DNA synthesis. For this work later, both Howard Temin and David Baltimore were awarded Nobel Prize. The enzyme was initially called 'RNA dependent DNA polymerase', but was later called 'RNA

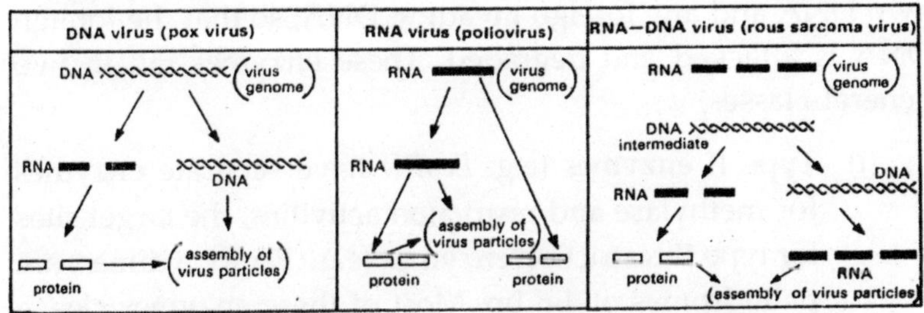

Three kinds of viruses, grouped on the basis of genetic material and mode of replication (a) DNA viruses, (b) RNA viruses and (c) RNA:DNA viruses.

directed DNA polymerase', since the purified enzyme was capable of utilizing a variety of templates including synthetic and natural DNA, RNA and RNA-DNA hybrids. The enzyme subsequently became popularly known as reverse transcriptase. This enzyme resembles other DNA polymerases and was detected in human tumour cells, initially suggesting its role in cancerous growth of cells. However, later this enzyme was found in the normal cells also, thus reducing the initial excitement about this enzyme being the cause of cancer in humans.

Although the discovery of RNA directed DNA polymerase or reverse transcriptase did not help in providing a solution to the problem of human cancer, the enzyme is now being extensively utilized for molecular biology studies. The most important of these uses is the synthesis of complementary DNA (cDNA) from mRNA, leading to the preparation of cDNA libraries for the study of genes expressed in specific tissues. A large number of such cDNA libraries have been prepared in a variety of organisms, both plants and animals, and have facilitated isolation of genes.

DNA Modification and DNA Restriction : Besides DNA replication apparatus, other enzyme systems (methylases) are found in cells that may modify DNA through methylation of adenine or cytosine in a very specific manner. There are also restriction enzymes in a cell that can distinguish between its own DNA and any foreign invading DNA, so that the foreign DNA is attacked and degraded. These enzymes fall in two general classes,

(i) Type II enzymes (e.g. *EcoRI)* have separate enzymes for methylase and restriction activities; the target sites for type II restriction enzymes (e.g. .EcoRI, a dimer) are palindromes of 4-6 bp. Most of these enzymes cleave DNA at an unmethylated site; some of them create staggered cuts, while others create blunt ends. A methylase (usually a monomer as in *EcoRI* system) adds one methyl group at a time and dissociates. Another round of binding and methylation is needed for addition of another methyl group. The uses of some of these restriction enzymes are discussed later, since they are extensively used in recombinant DNA technology,

(ii) Type I and Type III enzymes are bi-functional, having both methylase and restriction activities; a type I enzyme (e.g. *EcoK, EcoB)* has three subunits R, M and S (R for restriction; M for methylation and S for specific target site), but type III (e.g. *EcoP1, EcoP15)* has only two subunits (R for restriction and MS for methylation and recognition). The properties of these three types of enzymes are summarized in Table.

DNA Repair

Both in prokaryotic and eukaryotic cells, there are repair enzyme systems to deal with DNA, damage, which is caused rather frequently. The changes in DNA leading to damage are broadly divided into two general classes : (i) Single base changes may be caused by conversion of one base to another (e.g. deamination of 5-methylcytosine to thymine). These are corrected through DNA replication leading to changes in DNA

Three Types of DNA Restriction / Modification Enzymes

Property	*Type II enzyme*	*Type III enzyme*	*Type I enzyme*
1. Protein structure	Separate endonuclease and methylase	Bifunctional; 2 subtfnits	Bifunctional; 3 subunits
2. Recognition site	Short sequence (4-8 bp), palindromic	Asymmetric sequence (5-7 bp)	Bipartite; ± asymmetrical (e.g. TGAN$_8$ TGCT)
3. Cleavage site	Same as recognition site or close to it	24-26 bp downstrem of recognition site	> 100 bp from recognition site
4. Restriction & methylation	Separate reactions	Simultaneous	Mutually exclusive
5. ATP needed or not	No	Yes	Yes

sequences,(ii) Structural distortions may result from a single strand nick, removal of a base or introduction of a covalent link between bases of same or different strands (e.g. formation of thymine dimer due to UV).

A cell may have several systems to deal with DNA damage. These systems include the following : (i) Direct repair involves reversal of the damage; for instance, photoreactivation of pyrimidine dimer involves reversal of covalent bonds giving the original structure (no fresh DNA synthesis is involved). (ii) Excision repair involves recognition of damaged or altered base(s) followed by excision of a sequence including damaged bases. This is achieved by an endonuclease, or an exonuclease. A new stretch of DNA is then synthesized to replace the excised material. (iii) Mismatch repair involves correction of mismatches or pairing between bases which are not complementary. Mismatches may arise either (a) during replication or (b) due to base conversion (e.g. deamination or oxidative damage) and are corrected by a process described as error correction during DNA replication. Error correction involves a process that distinguishes between 'new' and 'old' strands (due to methylation), so that the error in the newly synthesized strand is corrected.

Direct Repair : As mentioned above, direct repair involves reversal of DNA damage. In other words, abnormal chemical bonds between bases or between a nucleotide and an abnormal substituent are broken down, thus restoring the original DNA structure. The enzymes currently known to catalyze this DNA repair include the following : (i) DNA photolyase repairs cyclobutane pyrimidine dimers (CPD), induced by UV. The enzyme splits the cyclobutane ring, only on activation by light. This enzyme is found in several bacteria (including *E. coli)* and some mammals, but is absent in bacteria like *B. subtilis* and in placental mammals. (ii) 6-4 photoproduct photolyase deals with the DNA damage involving the formation of '6-4 photoproduct' due to UV. The 6-4 photoproduct is formed firstly due to C4-C6 bond formation between adjacent pyrimidines and secondly, due to migration of a substituent available at C-4 position of one pyrimidine to the C-6 position of adjoining pyrimidine. The enzyme reverses both these anomalies and restores the normal bases: (iii) Spore photoproduct lyase repairs the lesion caused by UV in *B. subtilis* spores, where a spore photoproduct (5-thyminyl-5,6 dihydrothymine) is produced instead of cyclobutane dimers on UV irradiation. The enzyme breaks the C-C bond between the two thymines in a light-independent reaction, (iv) O^6-methylguanine DNA methyl transferase (MGMT) is a common enzyme found in all species tested. The enzyme transfers the methyl group of O^6 -methylguanine (less efficiently also the other alkyl groups at this position) to a cystene residue on the enzyme. O^6-methylation of DNA is a common effect of several simple alkylating agents including atmospheric methyl halides and products of nitrite metabolism, causing DNA damage.

Excision Repair Systems in E. Coli : Excision repair has been classified into the two types on the basis of the nature of excised products, which can either be free bases as in base excision repair (BER), or one or more nucleotides, as in nucleotide excision repair (NER). The lesions involving non-helix distortions (e.g. alkylating bases) are repaired by BER, and the lesions that involve distortion of the helical structure of DNA are repaired by NER.

Base excision repair (BER). BER is always initiated by a DNA glycosylase. A number of these glycosylases are known, depending upon the substrate used by each of them. These glycosylases hydrolyse N-glycosyl bonds linking bases to the sugar-phosphate backbone of the DNA, leaving sites that are abasic (without bases). These basic sites are then acted upon by an apurinic/ apyrimidinic (AP) endonuclease and an AP lyase, so that sugar-phosphate group is removed by hydrolysis of 5′phosphodiester bond. This leaves suitable template-primers for DNA polymerase to synthesize DNA and fill the gap. This results in a repair patch, which can be as small as single nucleotide. Some of the lesions in DNA, that are excised by BER, include uracil, thymine glycols, thymine hydrates, N3-MeAde and 8-oxo-guanine.

Nucleotide excision repair (NER). NER is initiated by endonucleolytic excisions, either only on one side of the lesion (described as endonuclease - exonuclease repair) or on each of the two sides of the lesion (described as excision nuclease or excinuclease repair). In the endonuclease-exonuclease repair, an endonuclease first makes an incision at either the 5' or the 3' end of the lesion. The exonuclease then digests segment of the strand involving the lesion. The gap is later filled by DNA synthesis and ligation. In excinuclease repair, incisions are made on either side of the lesion by endonuclease activities, giving fragments that are 12-13 nucleotides long in prokaryotes (8th bond on 5' side, and 4th or 5th on 3′side of the lesion) and 24-32 nucleotides long in eukaryotes (20th-25th bond on 5' side and 3rd-8th bond on 3′side of the lesion).

It is thus apparent that the distances of the incisions are located at rather precise distances from the damaged base. The enzyme system is called excision nuclease or excinuclease, and in *E. coli* it is due to combined action of three subunits [UvrA, UvrB, UvrC; also described as (A)BC excinuclease]. After the dual incisions, a gap is formed, to which repair proteins remain associated . These bound repair proteins are dissociated later due to association of replication proteins, which take part in repair synthesis of DNA to fill the gap, followed by ligation.

Although excision repair systems, discussed above, vary in their specificity, the main pathway involves the following three steps: (i) in incision step, an endonuclease cleaves the DNA strand on one or both sides of the damage (7 nucleotides from 5' side and 3 nucleotides from the 3' side); (ii) in excision step, a 5'-3' exonuclease (DNA polymerase I) removes a stretch of the damaged strand (20 nucleotides); (iii) in synthesis step, the single stranded region of DNA is used as a template for synthesis of DNA by DNA polymerase I to replace the excised sequence.

Excision repair, as above, involves different , lengths of DNA and is described as (i) very short patch repair (VSP), (ii) short patch repair and (iii) long patch repair. While VSP deals with single base mismatches, the other two deal with more extensive damages, which are repaired through *uvr* genes *(uvrA, B, C, D)*, coding for components of a repair endonuclease. UvrA and UvrB exhibit helicase activity ; UvrD (helicase II) is involved in the disassembly of post-excision complexes.

Mismatch repair systems in *E. coli mut HLS* pathway. When there is mismatch in a base pair as in a change from GC to GT, then theoretically it may repair to give rise to either wild type (GC) or to a mutant type (AT). Therefore, the repair system has to distinguish between old and new strands and repair only the new strand to restore the wild type. This requires four proteins, namely MutL, MutS, MutU (UvrD) and MutH coded in *E. coli* respectively by genes *mutL, mutS, mutU (UvrD)* and *mutH*. These *mut* genes are mutator loci, where a mutation leads to increased frequency of spontaneous mutations.

It has been shown that a methylase (dam) coded by *dam* gene, brings about methylation in adenine of sequence GATC on both strands of DNA. Following replication, one strand remains methylated (only at A of GATC sequence) and other remains unmethylated till methylase acts on this new strand to bring about methylation. During this transition period, the unmethylated GATC allows recognition of mismatch by MutL, and binding of MutS to mismatch. MutU helps in unwinding the single strands stabilised by SSB protein and MutH cleaves the newly synthesized strand. Two models for repair have been suggested which are shown in Figure. In one model

MutH cleaves DNA at two GATC sequences flanking the mismatch, while in the other model the cleavage occurs on one side of mismatch at GATC sequence and on the other side at the mismatch itself.

VSP and mutY pathways. The *mutHLS* pathway of *E.coli* described above corrects mismatches, which arise during DNA replication. There are two other pathways to correct mismaches arising due to action of chemical agents : (a) VSP corrects G-T mismatches (to GC pairs) created by deamination of 5-methylcytosine residues and also corrects G-U mismatches (though less efficiently). The pathway involves participation of the following five proteins (i) Vsr protein nicks DNA strand at a position 5' to the T and is stimulated by (ii) MutS and (iii) MutL; (iv) DNA polymerase I; (v)DNA ligase. (b) The *mutY* gene product corrects mismatches resulting from oxidative damage to G residues, which forms 8-oxo-dGTP, that may be degraded by MutT or removed from DNA by MutM. (8-oxo-G)-A mispair may change G-C to T-A, if not repaired by Mut-Y, with the help of DNA polymerase I and DNA ligase.

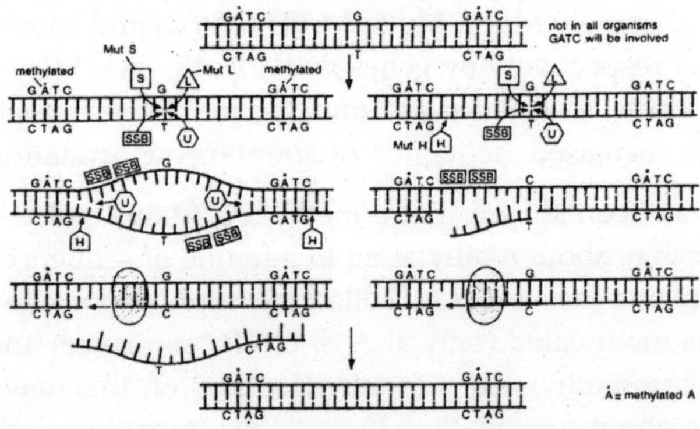

Two models for the mechanism of mismatch repair. In both models proteins called MutL and MutS interact with mismatch site (G—T) and a protein called MutH cleaves the newly synthesized strand. The repair apparatus distinguishes the parental strand from the new one by means of methyl groups (black dots) within the parental GA—TC sequences. The strands surrounding the mismatch are separated with the help of protein MutU and stabilized by SSB. (For details of two models, see text).

MutY also corrects A-G and A-C mispairs, though less efficiently.

Mismatch repair in Eukaryotes (Yeast and Humans) : Eukaryotic mismatch repair genes that are homologous to the corresponding genes in *E. coli,* have been identified in budding yeast *(S. cerevisiae)* and human cells. For instance, in yeast, five PMS genes *(PMS1, PMS2, PMS3, PMS5, PMS6)* have been identified that affect mismatch repair. Of these genes, *PMS1* was shown to be homologous to *mutL* gene of *E. coli.* Six other genes *(MSH1* to *MSH6)* were found to be homologous to *mutS,* of which *MSH2* is the most important for repair of nuclear DNA, while *MSH1* is important for repair of mtDNA.

At least four human genes *(hMSH2, hMLH1, hPMS1, hPMS2)* that have also been shown to be involved in mismatch repair, are linked to hereditary nonpolyposis colon cancer (HNPCC). DNA isolated from colorectal tumours and from HNPCC samples also exhibited microsatellite instability (microsatellites are mono-, di-, tri- and tetra-nucleotide repeats dispersed throughout the genome) involving expansion or deletion of (CA) repeats in the tumours. This suggested that the gene *hMSH2* of the mismatch repair system in humans is also involved in the maintenance of microsatellites and that a mutation in this gene causes HNPCC.

Relationship between NER and Transcription : In prokaryotes, as well as in eukaryotes, it has been shown that the nucleotide excision repair (NER) of template strand of the transcriptionally active DNA is faster than that of its complementary strand (also called coding strand). In prokaryotes, the subunits of the transcription factor TFIIH are required for NER, so that the factor TFIIH has a dual function of transcription and repair.

Recombinational Repair (Dimer Tolerance) Involves DNA Replication in E. Coli : Dimers, whenever formed (due to UV) in DNA, are reversed either by photoreactivation, due to

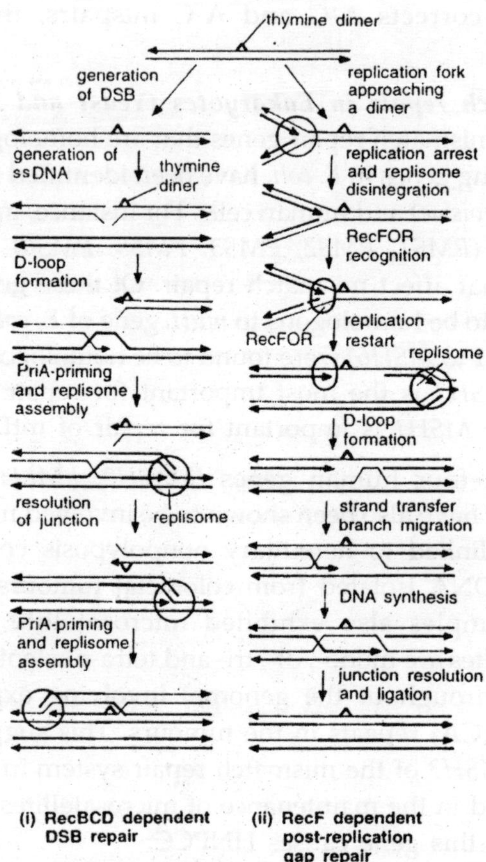

Two different mechanisms of DNA replication for recombinational repair in E. coli. (see text for details).

photolyase (direct repair) or removed by UvrABC catalyzed excision repair. However, some of the dimers still remain and are tolerated by the replication machinery. Although a modified DNA polymerase may replicate DNA, downstream of the dimers, by a mechanism called translation replication, major mechanisms of dimer tolerance involve the following two recombinational repair activities:(i) RecBCD- dependent double-stranded break (DSB) repair pathway. This pathway involves extensive DNA replication, which initiates at a D-loop, which is a recombination intermediate. (ii) RecF-dependent post replication gap repair.

In this pathway, RecFOR complex (RecF, RecO, RecR proteins) previously believed to be recombination proteins, were shown in 1997, to assist in replisome reassembly after UV irradiation. These two mechanisms of recombinational repair are shown in Figure. In both the above recombinational repair activities, newly synthesized DNA, after UV irradiation, contains single stranded (ss) gaps opposite the dimers. These gaps are later filled by a mechanism that requires homologous recombination.

An SOS Repair System in E. Coli : In *E. coli,* many treatments causing DNA damage or inhibition of DNA replication, induce a complex series of phenotypic changes described as SOS response. The SOS response (increased capacity to repair damaged DNA) is initiated by interaction of RecA protein with LexA repressor. Following steps are involved : (i) damaging treatment activates RecA protein (a protease), through common intermediates (small DNA molecules or single stranded region); (ii) RecA protease interacts with LexA repressor (coded by *lexA* gene) and due to proteolytic action of RecA on LexA, all operons to which LexA is bound are induced (Fig.); these may include a number of operons/genes (~30) called *din* (damage inducible) genes, such as *uvrA, uvrB, uvrC, polB, umuDC* and *hinA*. These genes each has a SOS box, 20 bp long having a 8 bp long consensus sequence. Since LexA represses its own synthesis and that of RecA, its proteolytic degradation leads to synthesis of both LexA and RecA proteins. However, when DNA damaging signal is removed, RecA is not induced and LexA accumulates, which represses all other genes belonging to SOS system.

DNA Repair and Genetic Diseases in Humans : Several genetic diseases in human beings are characterized by cellular hypersensitivity to DNA damage. These diseases also cause increased risk of cancer. Four such recessively inherited diseases include the following : (i) Fanconi anaemia (FA); (ii) Xeroderma pigmentosum (XP); (iii) Ataxia telangiectasia (AT) and (iv) Bloom's syndrome (BS); cells from patients with the first three diseases are uniquely sensitive to DNA damage by DNA

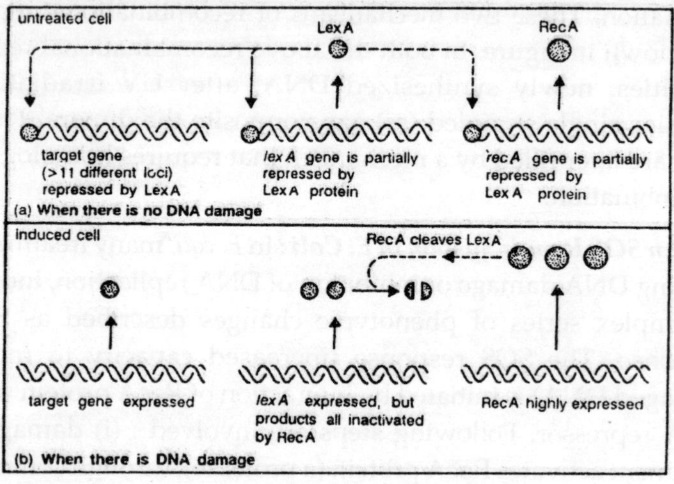

The SOS DNA repair system in E. coli.

Eukaryotic Nucleotide-excision-repair Genes and Mutants.

Human mutant (and gene)	*Rodent mutant*	*Yeast gene*
ATM *(XPAC)*	*?*	*RAD 14*
XP-B (XPBC)	*ERCC3*	*RAD25 (SSL2)*
XP-C (XPCC)	*?*	*RAD4?*
XP-D (XPDC)	*ERCC2*	*RAD3*
XP-E	*?*	*?*
XP-F	*?*	*?*
XP-G (XPGC)	*ERCC5*	*RAD2*
CS-A	*?*	*?*
CS-B (CSBC)	*ERCC6*	*?*
TTD1BR	*?*	*?*

Contd...

Human mutant (and gene)	*Rodent mutant*	*Yeast gene*
?	*ERCC1*	*RAD10*
?	?	*RAD1*
?	?	*RAD7*
?	?	*RAD16*

cross-linking agents, ultraviolet light and ionizing radiations respectively, those with BS are moderately hypersensitive to several agents causing different types of DNA damage. Genes responsible for two of these diseases (FA, XP) were identified by complementation of hypersensitive cell lines using DNA expression cloning technology. The XP DNA products show significant similarity to the products of the *RAD* DNA repair gene of budding yeast, and *ERCC* genes of rodents (Table), which are involved in the nucleotide excision repair showing that the essential enzymes involved in DNA metabolism have been conserved through evolution. In future, more work on the molecular mechanism of these and other diseases involving DNA damage will be done.

6

Immunotechnology

For human health care, the subject of the production and use of antibodies has become a very important area of research, not only for academic purposes but also for its relevance to industrial growth for production of vaccines and drugs for diagnosis and therapeutic uses.

For purposes of research and use in medicine, antibodies are often isolated by sacrificing animals after hyperimmunizing them with antigens. From such hyperimmunized animals, the blood serum may be taken and antibodies may be isolated from this serum. However, antibodies, whenever separated from serum after induction due to antigen, are usually heterogeneous, because the cells keep on producing a variety of antibodies. Consequently, the antibody preparation made from serum will not have the desired specificity and, therefore, cannot be utilized for diagnosis or screening. Monoclonal antibodies (mAbs), on the other hand, are homogeneous immunological reagents of defined specificity, so that these can be utilized for diagnosis and screening with earn and certainty. The discovery, production, genetic manipulation and uses of these monoclonal antibodies will be described in this chapter.

Making Monoclonal Antibodies by Hybridoma Technology

Basic Hybridoma Technology : Monoclonal antibodies can be produced in specialized cells through a technique now popularly known as hybridoma technology. TUB technology was discovered in 1975 by two scientists, Georges Kohler of West Germany and Cesal Milstein of Argentina, who jointly with Niels Jerne of Denmark were awarded the 1984 Nobel

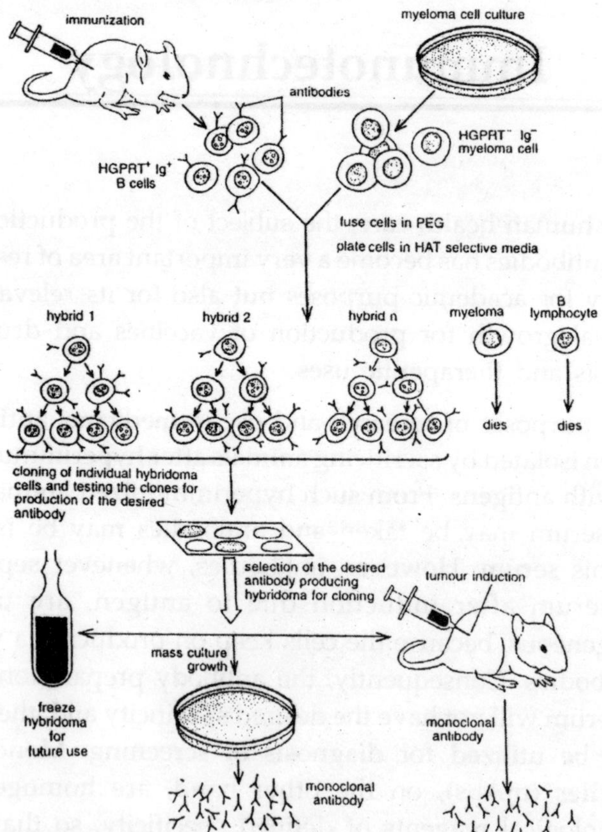

Procedure for making and selecting hybridomas for the production of monoclonal antibodies.

Prize for Physiology and Medicine. The term hybridoma is applied to fused cells resulting due to fusion of following two types of cells : (i) an antibody producing lymphocyte cell (e.g. a spleen cell of mouse immunized with red blood cells from sheep) and (ii) a single myeloma cell (bone marrow tumour cell) which is capable of multiplying indefinitely. These fused hybrid cells or hybridoma have the antibody-producing capability inherited from lymphocytes and have the ability to grow continuously (immortal) like malignant cancer cells. Following steps are involved in the production of monoclonal antibodies using hybridoma technology (Fig.) :

(i) Immunize a rabbit through repeated injection of a specific antigen for the production of specific antibody, facilitated due to proliferation of the desired B-cells.

(ii) Produce tumors in a mouse or a rabbit,

(iii) From the above two types of animals, culture separately spleen cells (spleen cells are rich in B cells and T-cells) that produce specific antibodies, and myeloma cells that produce tumors (the myeloma cell line used, is unusual in two ways; it has stopped synthesizing antibodies and it is a mutant called HGPRT" that cannot synthesize the enzyme hypoxanthine-guanine phosphoribosyl transferase or HGPRT).

(iv) Induce fusion of spleen cells to myeloma cells, using polyethylene glycol (PEG), to produce hybridoma; the hybrid cells are grown in selective hypoxanthine aminopterin thymidine (HAT) medium. HAT medium contains a drug aminopterin, which blocks one pathway for nucleotide synthesis, making the cells dependent on another pathway that needs HGPRT enzyme absent in myelome cells.

Therefore, myeloma cells, which do not fuse with B-cells will die since they are HGPRT". B-cells that do not fuse will also die because they lack tumorigenic property of immortal growth. Therefore, HAT medium allows selection of hybridoma cells, which inherit HGPRT gene from B-cells and tumorigenic property from myeloma cells.

(v) Select the desired hybridoma for cloning and antibody production; this is facilitated by preparing single cell colonies that will grow and can be used for screening of antibody-producing hybridomas; only one in several hundred cell hybrids will produce antibodies of the desired specificity;

(vi) Culture-selected hybridoma cells for the production of monoclonal antibodies in large quantity; these hybridoma cells may be frozen for future use and may also be injected in the body of an animal so that antibodies will be produced in the body and can be recovered later from the body fluid (Fig.).

Hybridoma Technology and Some Improvements : Considerable efforts during 1980-2000 were made to improve the yield of monoclonal antibodies using hybridoma technology. These efforts included the following : (i) As indicated above, cell fusions were facilitated through the use of polyethylene glycol (PEG), (ii) A continuous cell line (Sp 2/0) was used as a fusion partner for the antibody- producing B cells, (iii) Feeder layers consisting of extra cells to feed newly-formed hybridomas were used for optimal growth and hybridomij production. The most common feeder layers consisted of the following : (i) murine peritoneal cells; (ii) macrophages derived from mouse, raj or guinea pigs; (iii) extra non-immunized; spleen cells; (iv) human fibroblasts, peripheral blood monocytes or thymus cells.

These feeder cells; also had some limitations like depletion of nutrients meant for hybridoma and contamination, so that other sources off hybridoma growth factors (HGF) life interleukin-6 (IL-6) derived from human cells were used.

Bulk Culture of Hybridoma Cells and Serum Free Media : The culture media for animal cells can be either 'natural' or 'defined'. The defined media can be either serum-supplemented or serum-1 free. Serum is a highly complex and poorly-defined mixture of components like albumin, transferrin, lipoproteins and various hormones/I growth factors. Nevertheless, serum makes an essential component of media for culturing animal cells. The use of serum, however, leads to difficulties in purification of antibodies,

Positive and Negative Aspects the use of serum-free media in hybridoma cell culture and Production of monoclonal antibodies.

I. Positive Aspects

1. Greatly simplified purification of antibodies due to increased initial purity and absence of contaminating immunoglobulins.
2. Decreased variability of culture medium.
3. Reduced risk of infectious agents.
4. Fewer variables for quality control/quality assurance.
5. Increased control over bioreactor conditions.
6. Potential for increased antibody secretion.
7. Low or no dependence on animals.
8. Cost effective.
9. Overall enhanced efficiency.

II. Negative Aspects

1. Not all serum-free media are applicable to all cell lines.
2. Cells may not grow to as high densities and may be more fragile than cells in serum.
3. Media may take longer to prepare.

Further, it is an expensive technology for large-scale production of hybridoma cells for industrial production of monoclonal antibodies. In view of these difficulties, serum-free media are being increasingly used for culturing hybridoma cells. Advantages and disadvantages of the use of these serum-free media for culturing hybridoma cells are given in Table.

Antibodies and their Purification

Monoclonal antibodies may need to be purified before they are used for a variety of purposes. Before final purification, the cultures (whether cell cultures or bacterial cultures with cloned genes, as described later) may be subjected to cell fractionation for enrichment of the antibody protein. In *E. coli,* the antibodies may be secreted in the periplasm, which may be used for enrichment of antibody, so that further purification is simplified *(E. coli* cells can also be used for production of mAbs; see alter). Alternatively the antibodies may be purified from cell homogenates or cell debris obtained from the medium. Antibodies can be purified by any one of the following techniques (i) ion-exchange chromatography; (ii) antigen affinity chromatography.

Genetic Manipulations and Antibody Engineering : Although hybridoma technology enables us to obtain monoclonal -antibodies or homogeneous antibodies of predefined specificity, it can not allow production of novel antibodies of desired specificity. Such novel antibodies can be obtained by redesigning antibodies through the application of gene technology, an area of research which has revolutionized the potential of monoclonal antibodies. For application of gene technology, an initial but crucial problem was to find a convenient way for expression of the cloned antibody genes into proteins (antibodies). For this purpose expression could be obtained in (i) mammalian cells and (ii) bacterial cells, particularly *E. coli.* Initially, antibody genes were isolated from hybridomas, cloned into plasmid vectors and expressed as complete antibodies in mammalian cells. However, later, genes for different small fragments of antibodies, (e.g. Fv, Fab, Fc or even single chain Fv = scFv) could be cloned and expressed in bacteria (Fig.).

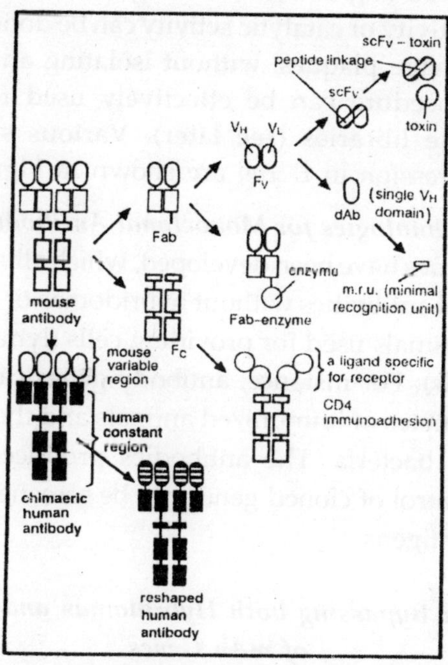

Fragments of Engineered Antibodies (see text for abbreviations used and for other details).

Expression in mammalian cells becomes important, when glycosylation of antibodies is desired. However, lack of glycosylation does not influence antigen-binding capacity of Fv or Fab fragments or of complete antibody, although it does affect some of the effector functions attributed to Fc fragment. Therefore, for industrial production of antibodies, glycosylation is immaterial and *E. coli* can be used as a useful expression system. Following are some of the advantages of the use of *E. coli* : (i) fast growth; (ii) simple fermentation to allow large-scale production in fermenters; (iii) fragments can be produced by Fv or Fab genes and not from complete antibody through proteolysis; these small fragments can be used for a variety of studies on structure (through NMR and crystallography) and function; (iv) antibodies in the form of bifunctional molecules can be produced (e.g. antigen binding + toxin); (v) screening for binding activity or catalytic activity can be done on bacterial colonies or phage plaques, without isolating antibodies; this screening procedure can be effectively used for screening antibody gene libraries (see later). Various strategies for antibody expression in *E. coli* are shown in Figure.

Other Technologies for Monoclonal Antibodies : In recent years, techniques have been developed, which allow production of monoclonal antibodies without hybridomas or even without the help of animals used for providing cells (lymphocytes and myeloma cells). For instance, antibody genes can be isolated from lymphocytes of immunized animals and then cloned and expressed in bacteria. The antibodies produced in bacteria under the control of cloned genes can be screened for binding to specific antigens.

Through Bypassing both Hybridomas and Cloning of mAb Genes

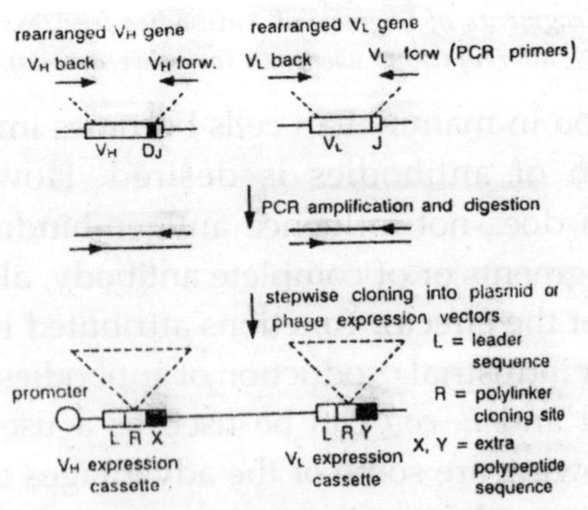

PCR cloning of rearranged Vh and Vl genes into expression vectors (see text for details).

Thus, while hybridoma technology can immortalize antibody-producing cells, gene technology immortalizes antibody producing genes. Computer graphic techniques are also being used to build specific antigen-binding sites in antibodies. Using this approach some designer antibodies of practical value have already been produced. In this latter strategy, genes are not really cloned from lymphocytes, but are instead designed from a repertoire of antibody genes available in a collection.

The V_H and V_L genes for antibodies can be amplified through polymerase chain reaction (PCR) using 'universal primers' (universal primers will carry conserved sequences for most antibodies). By building restriction sites in the above primers, the amplified V_H and V_L genes can also be cloned directly for expression in mammalian cells or bacteria (Fig.).

The raw material for PCR may be hybridomas or B cells, which may be homogeneous (if derived from single cells) or heterogeneous. In the latter case, a variety of V_H and V_L genes will be amplified and will combine at random to produce as many as 10^6 clones for antibody genes (from 1000 different V_H and 1000 different V_L genes).

These genes will be cloned in λ phage and their products (particularly Fab fragments) can be screened for antigen-binding activities. From such a large number of combinations in a combinatorial library, it is very difficult to recover the original pairs of V genes (e.g. $V_{Ha}.V_{LX}$ and $V_{HX}.V_{LX}$ are original pairs; $V_{Ha}.V_{Lx}$ is a new combination).

However, the complexity may be reduced by using antigen-selected B lymphocytes (filters coated with antigen can be used for screening). Different steps involved in cloning paired V_H V_L genes are outlined in Figure.

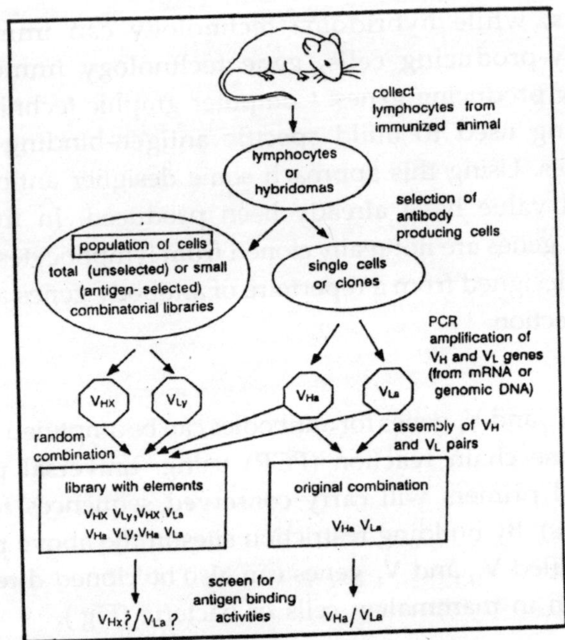

Strategies for cloning paired Vh and Vl genes from lymphocytes of an immunized animal.

Manufacture and Design of mAb Genes : The antigen binding sites of antibodies have been studied in some detail in recent years. This led to modelling of entirely new antibodies, sometimes for their use as enzymes. This modelling thro ugh computer graphics can be used for alteration of antibody genes or for synthesis of entirely new genes. These genes can be cloned and expressed in bacteria. The antibodies produced can be tested for their specificity and affinity for specific antigens.

Antibody Genes their Primary and Secondary Libraries : In this method, a repertoire of antibody genes can be prepared by using genes that can be obtained from a number of different sources eluding the following : (i) rearranged V genes from animals obtained through the use of PCR (with universal primers); (ii) new V genes obtained through gene conversion, a process adopted in birds; (iii) rearranged genes obtained from mRNA through reverse transcription; (iv) designing entirely new V genes or D segments. The next step is to allow the expression of library in bacteria and screen antibodies for antigen-binding activities. The screening can be done on membrane filters coated with antigen. In future, the screening procedures may be replaced by methods of selection. In either case the selected V_H and V_L genes can be subjected to mutations to increase the affinity of an antibody for a specific antigen. A variety of methods for the above strategy are being developed, so that monoclonal antibodies can now be produced without hybridomas and lymphocytes.

Humanized and Human Antibodies and their Preparation— Humanized Antibodies : Human antibodies are currently produced by the following methods : (i) fusion of mouse myeloma cells with human lymphocytes (blast cells in peripheral blood lymphocytes); (ii) immortalization of human cells by Epstein-Barr virus. Both the methods have limitations. Human-mouse hybrid cells have a tendency for preferential loss of human chromosomes, making them unstable. Similarly, Epstein-Barr virus does not allow preferential immortalization of blasts engaged in antibody response.

In view of these difficulties 'humanizing of rodent monoclonal antibodies' through genetic engineering is the most practical approach, which has been used for the production of mouse-human chimeric monoclonal antibodies to be used for tumour therapy or for manipulation of human immune system or against cell surface antigens. The humanized chimeric antibodies combine the rodent variable regions with the human constant (or constant + variable) framework regions. To further reduce the immunogenicity of rodent elements, humanized antibodies have been produced, which retain only the antigen-binding complementarity- determining regions (CDRs) from the rodent in association with human framework regions.

Human Antibodies **:** In early 1990s, two parallel technologies were developed to generate repertoires of fully human mAb (monoclonal antibodies). These repertoires, once generated can be used for selection of antibodies with specific binding sites. Following two technologies have been used (i) transgenic mice and (ii) phage antibody display.

Transgenic mice for human antibodies. The transgenic approach exploits the natural diversity/selection strategies used in mouse system and makes use of these strategies for production of human antibodies in transgenic mice. These strategies seem to work in a similar way in both mouse and humans, so that the human immunoglobin genes seem to undergo rearrangement and hypermutation in transgenical mice in the same way as in humans. In this approach mouse immunoglobin loci are disrupted and human transloci perform without any competition from the endogenous loci.

One of the initial problems in the production of transgenic mice with human loci was the large size of human immunoglobulin loci, which stretch over several megabases (thousands of kilobases). In view of this, only a portion of total V-gene repertoire was exploited to provide limited antibody diversity in the repertoires thus generated. Early studies actually used mini-loci, which contain only a small number of selected V-gene segments, ill association with almost normal number of diversity (D) and joining (J) segments. These segments provide enough diversity and further diversity can be added through somatic hypermutations. In recent years, however, the technique for transfer of yeast artificial chromosomes (YACs) into embryonic stem cells allowed the introduction of large segments of human immunoglobulin genes into mouse germ line leading to the production of transgenic mice producing antibody diversity of the order normally found in human beings.

In 1997, success was achieved to produce transgenic mice carrying long segments of immunoglobulin genes and producing human mAbs. From these successful reports, there seems little reason to prefer smaller loci. In future, transgenic mice will be produced, which will carry full complement of human variable and constant regions. Efforts are also being made to increase the affinity of the mAbs produced in transgenic mice. The advantage of transgenic approach over antibody phage display technology discussed below is that the high affinity mAbs can be directly selected using well established hybridoma technology. But phage display in much faster.

Phage antibody display for human antibodies. The phage display technology involves fusion of the DNA sequence coding for antibody variable (V) region, to the sequence coding for the amino terminus of the minor coal protein pill of the phage. A large collection of combinatorial repertoires of seauences coding for diverse antibodies needs to be created to be used for fusion with phage coat protein gene sequence. These sequences will have different antigen binding sites generated by a PCR-based amplification of V genes.

The fusion protein, when synthesized under the influence of this fused sequence in the phage, allows the antibody being presented on phage surface. From these phages with a variety of antibodies on their surface, one can select antigen- specific phage antibodies using the following strategy. The labelled antigen is immobilized on a surface and the surface is exposed to the phage with antibodies, so that the phage displaying the antigen- specific antibody will be retained on the surface coated with the antigen, and non-adherent phages displaying irrelevant antibodies will be washed away. Phages displaying bound antibodies can be recovered from the surface, and used for re-infecting the bacteria. These bacteria will be grown for further enrichment of antigen- specific antibodies.

The availability of phage display technology in early 1960's revolutionized the generation and engineering of monoclonal antibodies. In view of this, mAbs for chosen target antigens can now be produced without the use of any animals or hybridomas, bypassing completely the immune system. Single-pot phage libraries can now be prepared and the affinity of antibodies can be improved to picomolar level. The use of phage display will gradually replace hybridoma technology in all its applications, both in research and diagnostic industry.

'Single-pot' antibody libraries. Although antigen-biased phage antibody repertoires can be prepared using V-genes from immune sources, it is always better to produce antibodies from universal antigen-unbiased libraries. Two types of such universal libraries (called 'single pot' libraries) have been described : (i) naive (not immune) antibody library, and (ii) synthetic antibody library. The naive antibody library is based on cloning a repertoire of an array of rearranged genes for IgM antibodies (IgM appears early in response and is later replaced by IgG). These genes are derived as PCR products amplified from cDNA reverse transcribed from IgM mRNA, which can be isolated from peripheral blood lymphocytes (PBLs), bone marrow or spleen cells or unimmunized human donors. The primers for PCR may be designed from sequences of the framework regions (Frs) 1 to 4. These cassettes are cloned in phage or phagemid vectors for expression and display of antibodies.

In 1991, first such 'single-pot repertoires' were made from PBLs of two healthy volunteers and it contained 3×10^7 clones, form which antibodies against at least 25 antigens were selected. Much larger scFr repertoires (10^9-10^{10} clones) were made in 1996 by brute force cloning. Synthetic antibody libraries, on the other hand, were made from unrearranged V-gene segments on the basis of six CDRs that shape the antigen-binding site of the antibody. The areas of greatest natural sequence diversity are chosen for these synthetic libraries and, therefore, CDR3 of the heavy chain became the region of choice, because the other five CDRs have limited variation. Libraries are now available having 95% of the segments used in natural human repertoire.

Monoclonal Antibodies and their Uses

Diagnosis, Purification and Therapy : Monoclonal antibodies or specific antibodies, are now an essential tool of much biomedical research and are of great commercial and medical value. For instance, ABO blood groups could be earlier identified with the help of human sera-carrying antibodies of known specificity. These human sera in U.K. have been replaced by monoclonal antibodies produced by hybridomas, for the identification of ABO blood groups. Thus, the diagnostic and screening value of the monoclonal antibodies through serological tests has been demonstrated. Besides the use of monoclonals in identification of blood groups in UK (UK blood typing), following three uses for monoclonals are described, although, only the first two of these make a definite market at present : (i) diagnosis (including ELISA test for detection of viruses and imaging), (ii) immuno-purification; (iii) therapy.

Diagrosis and Monoclonal Antibodies : In diagnosis, pregnancy can be detected by assaying of hormones with monoclonals. Similarly, pathogens can be detected in a few hours sparing several days of culturing of cells earlier needed. Radio immunodetection is also widely used. $(Fab')_2$ and Fab fragments are preferred for imaging, because both targeting and blood clearance are rapid. Tumours as small as 0.5 cm, which are missed by other radiological methods, can be detected by radio-labelled antibodies or Fab fragments.

Purification and use of Monoclonal Antibodies : Immunopurification involves separation of one substance from a mixture of very similar molecules. For instance, individual interferons could be purified using monoclonal antibodies and could be used for inactivating T lymphocytes responsible for rejection of organ transplants. Removal of tumour cells from bone marrow is another therapeutic use of monoclonal antibodies.

Use of Monoclonal Antibodies for Therapy : For therapeutic uses, monoclonal antibodies are so designed that they will neutralize the reaction or response by one defined antigen but will preserve the reaction of all other antigens. Several antigens of T cell receptor complex, including CD3, CD4 and CD8 have been the targets of specific antibodies for therapy. Most-widely used monoclonal antibody is OKT3, which has been licensed for clinical use, particularly for the treatment of acute renal allograft rejections. Monoclonal antibodies have also been used for treatment of patients with (i) malignant leukemic cells, (ii) B-cell lymphomas, and (iii) a variety of allograft rejections after transplantation. However, in several cases of therapy by monoclonal antibodies, there are undesirable side effects like high fever, vomiting, diarrhoea and respiratory distress; sometimes it also leads to broad immuno-suppression leading to increased incidence of infections. For many human diseases as above, murine (of mouse) antibodies are utilized. Their effectiveness is, however, limited due to their short survival time in humans and relatively low cytocidal effect. For this reason, humanized antibodies are being produced by genetic manipulations and in future these will be increasingly utilized for therapy, 'Monoclonal antibodies-cytotoxic agent conjugates' called 'immunotoxins' have also been designed as carriers of cytotoxic substances to the target cells.

'Radiolabelled monoclonal antibodies' have also been developed for delivery of a oytotoxic effector to target cells. For therapy, one advantage of using radiolabelled antibodies, is that they can kill cells from a distance and thus can also kill cells adjacent to antigen expressing cells. Further, the radiolabelled antibody need not be internalized to kill the tumour cells. For this purpose, both β emitting (e.g., ^{731}I, ^{90}Y, ^{186}Re, ^{188}Re and ^{76}Cu) and a emitting radionuclides (^{211}At, ^{212}Pb, and ^{212}Bi) have been used for patients with hepatoma, HTLV-1 (human T-cell leukemia/lymphoma virus -1), ATL (adult t-cell leukemia) and those with a variety of allografts, etc.

Allergic Diseases and Anit-IgE Therapy : Most allergic diseases are caused by IgE-mediated hypersensitive reactions. IgE is present in the serum in minute quantities and is produced by IgE-secreting plasma cells, which express antibodies on their surface. Patients suffering with allergy, produce increased amount of IgE, specific for innocuous antigens, to which the patients are sensitive. In an allergic reaction, the inhaled or ingested allergens bind to IgE on mast cells or basophils, help cross-linking of IgE molecules and aggregation of IgE receptors, thus triggering the cells to release histamine or other mediators of symptomatic allergic response.

The available drugs for allergy (antihistamines, corticosteroids and bronchodilators) treat symptoms and concomitant inflammatory reactions, rather than treating the process. Therefore, efforts are being made to treat the process by either inhibiting the synthesis of IgE or neutralizing free IgE in the cell. Humanized anti-IgE monoclonal antibodies with unique specificity for treating IgE-mediated diseases have actually been developed at the commercial scale. Phase I, phase II and phase III treatments with anti-IgE antibodies have already been conducted with positive results. Food (e.g. peanut) - induced allergy has also been treated. These trials suggested that anti-IgE antibodies are safe and efficacious for treatment of patients suffering with IgE-mediated allergy and other diseases.

Vaccine Production and Monoclonal Antibodies : Antibodies have also been used to immunize against certain diseases in humans and cattles. The most promising outcome in this area is the prospect of developing antimalarial vaccine in the near future. Monoclonal antibodies that inhibit the *in vitro* multiplication of *Plasmodium*, and the anti-gametocyte antibodies that inactivate male gametes have been developed. Monoclonal antibodies that destroy merozoite infected red blood cells have also been developed now. Such antibodies may prove useful as vaccines.

Use of Monoclonal Antibodies as Enzymes (Abzymes) : The production of novel enzymes through enzyme engineering has been a thrust area of biotechnology research consult author's book *"Biotechnology and Genomics"*. Antibodies can also be modified to be used as enzymes and are then described as abzymes. The antibodies may often bind specific ligands (haptens), but may not carry out chemical reactions. By modifying these ligands, antibodies may be generated that will catalyse specific reactions just like enzymes. Production of these abzymes is based on the following two principles : (i) enzymes work by binding the transition state of a reactant better than the ground state : (ii) antibodies, which bind to specific small molecules, can be produced by coupling this small molecule to a protein carrier and using this protein for immunizing experimental animals. If this small molecule is a transition state analogue (molecules that will mimic the shape and electronic configuration of transition state of reactant), then the antibodies that were produced to bind to this molecule will function as an enzyme towards the substrate of this reaction. Considerable success in the production of abzymes catalyzing following reactions has been achieved : (i) acyl transfer reaction, (ii) carbon-carbon bond formation, (iii) carbon-carbon bond cleavage. A number of monoclonal antibodies using the above approach have been produced to be used as abzymes. These abzymes were also shown to accelerate the individual reactions which they were supposed to catalyse.

Quantitation of Molecules and Purification through Monoclonal Antibodies : Antibodies have been used for the purification and quantitation of certain molecules present in trace amounts, such as hormones, cyclic nucleotides, polypeptides, enzymes, antigens, etc. The assay is extremely sensitive and quantities in as low as nano-or picomolar concentrations can be detected in small volumes (lml) of body fluids such as plasma, urine or cerebrospinal fluid. Following steps are involved in the purification and quantitation of enzymes and other proteinaceous molecules.

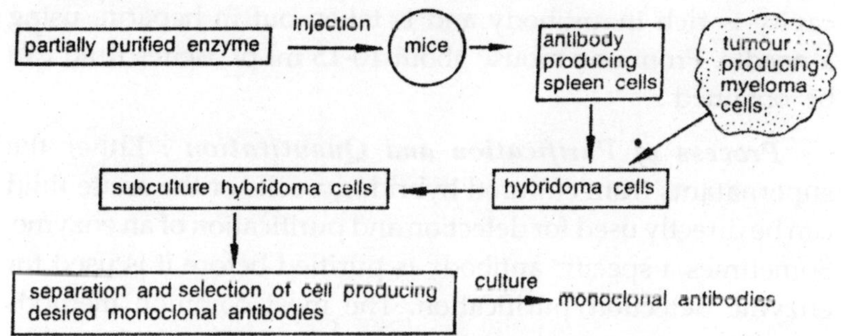

Monoclonal Antibodies and Steps of Production

How to Raise the Antibodies? : Monoclonal antibodies for specific enzymes or proteins are raised using traditional hybridoma technology. The enzyme is first extracted and partially purified using most of the conventional steps, such as extraction, ammonium sulfate precipitation and gel/column filtration. The preparation is diluted and emulsified with 'Freunds complete adjuvant'. Then it is injected intramuscularly into mice, to raise antibodies. Generally several doses, at intervals of a few days, are injected to boost the antibody production. The spleen cells from the mice are collected and fused with myeloma (tumor) cell lines derived from a rat. The resultant hybridoma cells are screened for their capacity to produce antibodies and secrete them out (Fig.). The hybridoma cells can be frozen for storage and can be thawed and recultured for the production of antibodies, whenever needed.

The amount of antibody production can be substantially increased by ascites fluid production method, which can replace hybridoma cell cultures. It has been shown that hybridomas in culture produce only up to 400 (ig/ml of specific antibody, while in ascites fluid, the amount of antibody is about 100-200 times more. In ascites fluid method, hybridoma is injected intraperitonealy into histocompatible mice to produce ascite cell tumours. One week to 10 days after injection, the peritoneal cavities in the mice appear swollen. The ascites fluid from the cavity is rich in antibody and is taken out in heparin, using a needle. From one mouse about 10-15 ml of ascites fluid can be obtained.

Process of Purification and Quantitation : Either the supernatants from cultured hybridoma cells or the ascite fluid can be directly used for detection and purification of an enzyme. Sometimes a specific antibody is purified before it is used for enzyme detection/purification. The most common methods for enzyme purification/ detection are : affinity chromatography, radioimmunoassay and immunoprecipitation.

(a) *Affinity Chromatography :* In affinity chromatography, the monoclonal antibody is generally immobilized on an insoluble matrix such as cross linked dextran or agarose beads, by a covalent linking agent (e.g. cyanogen bromide). The immobilized antibody is generally packed in a column, where its average concentration is 5-10mg antibody per ml of bed volume. Protein solution containing a specific enzyme is applied over this immobilized antibody. The specific enzyme binds with the antibody while other components do not, and thus the specific enzyme is separated. The enzyme can be recovered later by using suitable reagents. Antigens can also be purified using affinity chromatography with monoclonal antibodies. The antigen, which binds with the immobilized antibody, is eluted and recovered from the immobilized support by suitable washing procedures. Change in pH of the buffer often causes elution. Some important proteins such as alpha fetoprotein and leucocyte interferon are also purified using this procedure.

(b) *Radioimmunoassay* : Radioimmunoassay is an extremely sensitive method for the detection and quantitation of any substance that is antigenic and that can be labelled with a radioactive isotope. Basically the method depends upon the competition between labelled (known) and unlabelled (unknown) antigens for the same antibody. A known amount of labelled antigen (enzyme in this case), a known amount of specific antibody and an unknown amount of unlabelled antigen in the sample are allowed to reach equilibrium. By measuring the amount of labelled antigen bound to a known amount of antibody, a measure of antigen in unknown sample can be obtained. By suitably adjusting the amounts of antibody and labelled antigen, even small amount of antigen in an unknown material can be detected.

In order to measure the amount of labelled antigen attached to antibody, it is necessary to separate the antigen-antibody complexes from the mixture. By measuring the radioactivity, the percentage of labelled antigen bound to the antibody can be calculated. The concentration of an unknown (unlabelled) antigen can be determined by reference to a standard curve constructed from data obtained by allowing varying amounts of unlabelled antigen to compete (Fig.).

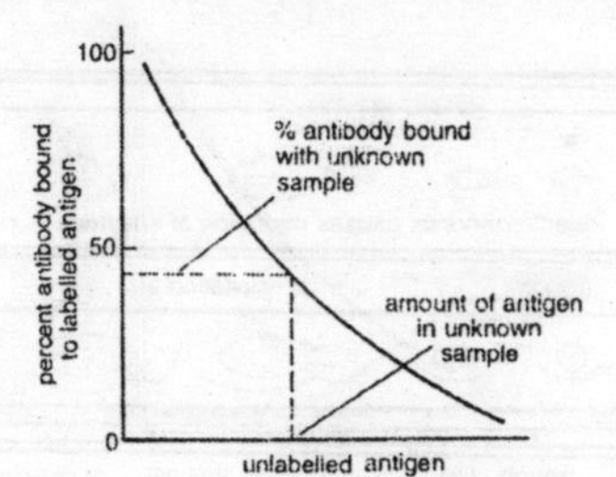

A good curve in use of radioimmunoassay for quantitative estimation of antigen in a sample.

(c) *Immunoprecipitation* : When correct conditions are employed, antigen and antibody react to form a precipitate. This precipitation phenomenon has been used to separate and purify enzymes and other antigenic substances in a variety of ways. One application of immuno- precipitation is immunoelectrophoresis. This is the most sensitive method for the detection of enzymes and antigens in a mixture. It involves combination of electrophoresis and gel diffusion. The mixture containing the enzyme (or antigen) is placed in a small well, made in a layer of agar placed on a glass sheet, or a microscopic slide. It is then subjected to electrophoresis by application of an electric current. The enzyme and other components of the mixture migrate through the agar at variable rates. Following electrophoresis, a trough is cut in the gel, parallel to the electrophoretic migration (Fig.). An antiserum (containing anitbody of that specific enzyme) or monoclonal antibody preparation is then placed in the trough. The reactants diffuse towards one another and form separate arcs of precipitate for each antigen-antibody complex. If only one monoclonal antibody preparation is used, then only one precipitation arc is formed.

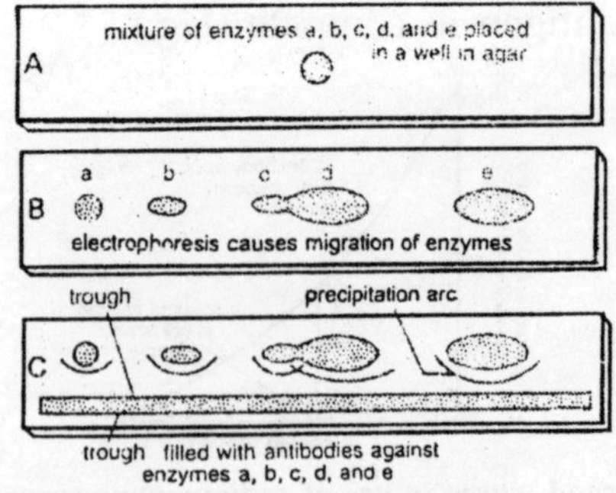

Detection and separation of enzymes in mixture and method used.

Monoclonal Antibodies for Cytogenetic Analysis : The technique of ELISA (enzyme linked immunosorbant assay), utilizing monoclonal antibodies, has also been used for cytogenetic analysis in wheat. Monoclonal antibodies (mAbs) specific to several chromosomes (e.g. IB, 6A and 6D) are already available and those for other chromosomes will become available in future, due to proteins coded by chromosomes of all the homoeologous groups (Table) Nullisomic and monosomic conditions for the chromosome IB and the absence of IBS in 1BL/1RS wheat-rye translocation (IBS carries gene for wheat gliadin protein) could be ascertained using the monoclonal antibody P24B, which is specific for IBS gliadins. Relative binding of P24B to 50% propanol extracts of half kernels could be used as a criterion for identification of corresponding nullisomics (showing no binding or very low binding) and monosomics (intermediate binding), because disomics show high level of binding.

Sometimes nullisomics and monosomics cannot be distinguished from each other with complete certainty, so that the results need to be confirmed through chromosome analysis. But mAbs can be used to enrich the number of aneuploids in a population, by rejecting seeds showing high level of reaction with mAb. The major advantage of using mAb is that a very large number of seeds can be examined non-destructively (only a part of seed is needed for mAb reaction) and the chromosome analysis can be greatly reduced, if not eliminated.

Homologous Groups with Chromosomes having Genes for Different Proteins

Class of Proteins	*Homoeologous groups having chromosomes with genes for proteins*
1. Endosperm proteins	1, 6
2. Amylase inhibitors	2, 3, 6
3. Albumins and globulins	3. 4, 5, 6 and 7

7

The T-cells

T-cells develop in the thymus and are responsible for cell-mediated immunity. These specialized cells react with fragments of forcign antigens that are associated with major histocompatibility complex or MHC molecules, and displayed on the surface of the other host cells infected by foreign agents like viruses. In this reaction, the host cells with fragments of foreign antigens on the surface are killed and eliminated before the virus inside these cells can be multiplied. In other cases, these T-cells may secrete chemical signals (lymphokines, interleukins or cytokines) that activate macrophages to destroy the invading microorganisms. Like B-cells which produce antibodies, T-cells are also antigen- specific and are equally important in defending vertebrates against infections. However, T-cells differ from B-cells in many ways (Table).

As shown in Table, the T-cells respond to the presence of peptide fragments on the surface of infected host cells. These peptide fragments are carried to the cell surface by special proteins called MHC molecules (Fig.), encoded by a complex of genes designated as *MHC*. The T-cells recognize foreign antigens associated with self-MHC molecules rather than foreign MHC molecules. The phenomenon is described as MHC restriction. The MHC molecules and their genes *(MHCs)* have also been an area of very active research during the last quarter of the 20th century. The 1996 Nobel prize in Medicine was actually awarded to Rolf Zinkernagel and Peter Doherty in recognition of their work involving 'MHC restriction' published in the year 1974.

Distinguishing Features of B Cells and T Cells of the Immune System.

B-cells	*T-cells*
1. B-cell secrete antibodies, which can act at distances far away; when antibodies are present on the surface of B-cells, they are called B-cell receptors.	1. T-cells act at short range, killing target cells; they do not secrete antibodies (but secrete interleukins); they instead have on their surface T-cell receptors having homology with antibodies.
2. B-cells can recognize free antigens in the circulating system of the body (not necessarily on the cell surface), including blood/ lymphatic system.	2. T-cells recognize antigens only on the cell surface and that too in the presence of major histocompatibility complex (MHC); they also recognize micro-organisms and foreign extracellular antigens ingested by the cell.
3. B-cells recognize intact antigens and not their fragments.	3. T-cell recognize peptide fragments of antigens that are partially degraded inside the cell and carried to the cell surface.

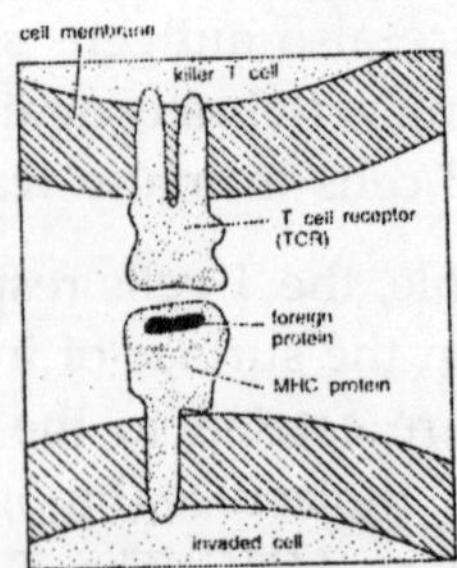

Reaction of T-cell, against a fragment of foreign protein, only when associated with a MHC molecule.

T-Cells : Kinds

There are mainly two types of T-cells : (i) cytotoxic T lymphocytes (CTLs) and (ii) helper T-cells. The former kill the cells that are infected with a virus or any other microorganism, and the latter send chemical signals to help stimulate the responses of other cells like macrophages, B-cells and CTLs. There is yet another type of T-cells called suppressor T-cells, which block B and T-cell responses, whenever so required. The helper T-cells are also of two types, Thl cells which activate macrohages and Th2 cells which stimulate B cells and eosinophils.

CD3 Complex and Receptors of T-cells : T-cells possess antibody like receptors called T-cell receptors, found on the surface of these T-cells to allow response through direct contact

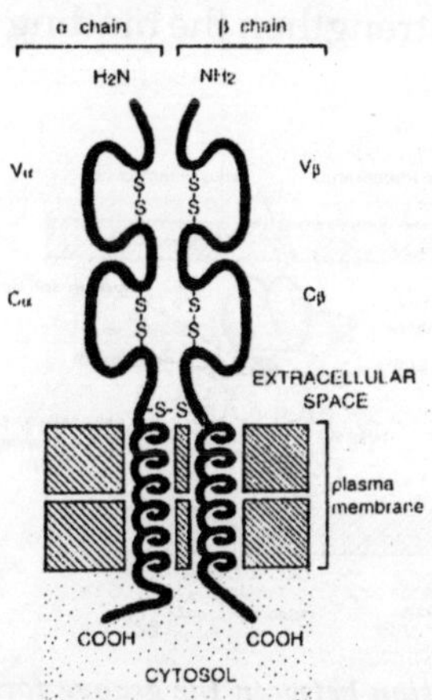

A T-cell receptor heterodimer, consisting of a and P polypeptide chains, joined by disulphide (S-S) bond (redrawn from Alberts el al. "Molecular Biology of the Cell", 1994).

with target cells. These receptors are encoded by genes that are assembled from multiple gene segments during T-cell development in the thymus, but are not secreted like antibodies, as in case of B-cells. Therefore, it was difficult to isolate them and they were identified for the first time as late as 1983. Each receptor (both in CTLs and helper T-cells) consists of two polypeptide chains (called α and (β) linked with disulphide bonds (Fig.). Early in development, some T-cells, found in epithelia, make a receptor heterodimer consisting of γ and δ chains instead of α and β. Each polypeptide consists of two Ig-like domains and resembles antibodies in having amino-terminal variable and carboxy-terminal constant regions. The diversity in T-cell receptors is achieved in the same manner as the antibody diversity is achieved in case of B-cells.

Antigen-driven somatic hypermutations, a characteristic of B-cells is, however, absent in T-cells and explains relatively low level of diversity (in comparison of antibody diversity in B-cells) in T-cell receptors. This low level of diversity is compensated by antigen non-specific cell-cell adhesion mechanisms, which strengthen the binding of T-cell to its target cell.

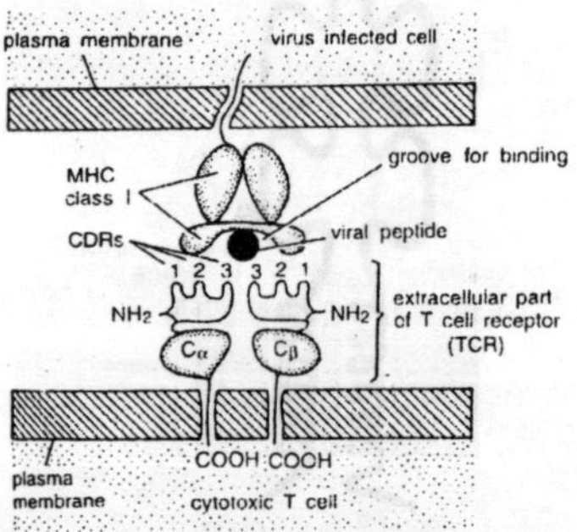

A model showing interaction between the groove formed by the (3 strands of MHC class I molecule with the CDR loops of T-cell receptor (TCR), as revealed by the three- dimensional structure of TCR reported in 1996.

The three-dimensional structure (TDS) of a T-cell receptor (TCR) with class I MHC molecule was also reported in 1996. It was shown in this study that a and p chains of the TCR, fold into a quaternary structure that resembles Fab region of antibodies. In this structure, the variable domains resemble antibody variable domains, which are described as complementarity determining regions or CDRs.

The TDS also revealed that the set of CDR loops align with the P strands, that form the groove of class I MHC molecule. This allows interlocking of TCR and MHC structures. It was also shown that four of the seven potential sites of glycosylation are indeed glycosylated.

A T-cell receptor is always associated with a set of transmembrane proteins, which are invariant. These invariant transmembrane proteins are described as CD3 complex. This complex helps in signal transduction, transferring information about extracellular binding events (binding of T-cell receptor with MHC molecule) to intracellular signals bringing activation of T cells. This signal transduction is brought about by activating one or more members of the Src family of tyrosine kinases (e.g. Fyn protein).

The kinases, in their turn bring about phosphorylation of various cellular proteins, including those of the CD3 complex and phospholipase C-y (PLC-y). These phosphorylated proteins then activate the inositol phospholipid signaling pathway. A comparison of T-cell receptors and B-cell receptors is shown in Figure.

In addition to T-cell receptors and CD3 complex, a variety of co-receptors also play an important role in activation of T-cells. These co-receptors are described later in the chapter. In this manner, T-cell receptor, co-receptors, CD3 complex and tyrosine kinases form a large signaling complex in the plasma membrane of the T-cell. This complex needs another, independent signaling pathway (described later) to activate the helper T-cell.

T-cell binding to the target cell is also facilitated by other accessory proteins like lymphocyte- function-associated protein (LFA-1), a member of the integrin family of cell-adhesion proteins. LFA-1 binds to a ligand, the intercellular adhesion molecule-1 (ICAM-1), which is displayed on target cell.

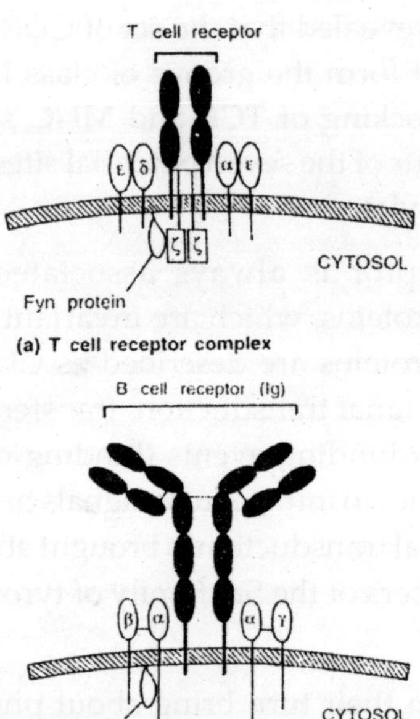

A comparison of the antigen receptors on T-cells and B-cells; B-cell receptors are actually antibodies on the surface of B-cells (redrawn from Alberts et al., "Molecular Biology of the cell", 1994).

Antigens to T-Cells and MHC Colecules : MHC molecules were first reported in organ transplantation reactions where graft rejection was witnessed as T-cell mediated immune response against foreign cell surface proteins called histocompatibility molecules (histo = tissue). The MHC family of proteins encoded by a cluster of genes called major histocompatibility complex or *MHC* are the most important of these surface proteins. They were initially called H-2 antigens (histocompatibility-2 antigens) in mice and HLA antigens (human-leucocyte-associated antigens) in humans (because these were first detected in leucocytes in humans). Following are some characteristic features of these MHC molecules : (i) T-cell mediated transplantation rejection is mainly due to MHC molecules; (ii) A large proportion of T-cells recognize foreign MHC molecules and not the foreign antigen, so that the antigen cannot cause a response, unless it is presented to the T cells by MHC molecule; only .001% of an individual's T-cells can respond to a viral antigen; (iii) there are as many as five loci in an individual for MHC molecules and as many as about 100 alleles at each locus, thus providing a very high degree of polymorphism and making it difficult that same MHC [molecules will be available in the donor and the recipient involved in organ transplantation (except in identical twins). These MHC (molecules not only help T-cells to reject foreign tissue invading the body, but also help them to focus on their own target cells that have foreign antigen on their surface.

Thus, although originally discovered as the molecules that lead to rejection of foreign grafts in organ transplantation, their major function, as later shown is to present the infected cells (with peptide fragments on their surface) to the T-cells for their destruction as a defense mechanism.

Difference Between Class I and Class II MHC Molecules.

Class I MHC	*Class II MHC*
1. It is found on all nucleated cells; is involved in binding peptide fragments from intracellular antigens (e.g. virus), and presenting the cells to cytotoxic T- cells.	1. It is found on specialized cells (e.g. B cells; thymus cells); is involved in binding extracellular or endocytosed peptide fragments and presenting them to helper T cells.
2. It consists of a single polymorphic transmembrane polypeptide chain called a and extracellular (consisting of three domains $\alpha_1, \alpha_2, \alpha_3$) invariant protein β_2 microglobulin.	2. It consists of two transmembrane polymorphic polypeptides α and β each consisting of two extracellular domains (α_1 and α_2 ; β_1 and β_2).
3. β_2 microglobulin is not encoded within *MHC;* it is not glycosylated and is non-covalently associated with three domains of *a.*	3. Both chains (α, β) are encoded within MHC; both are glycosylated and remain non-covalently associated.
4. α_3 , and β_2 are Ig-like.	4. α_2 and β_2 are Ig-like.

Contd...

Class I MHC	*Class II MHC*
5. Two outermost domains (α_1, β_2) are polymorphic and are meant for binding; the bound peptide is 8-10 amino acids long.	5. Two outermost domains (α_1, β_1) are polymorphic and form the groove for binding; bound peptide is 15-24 amino acids.
6. Genetic loci; *H-2K, H-2D, H-2L* in mice; *HLA-A, HLA-B, HLA-C* in humans.	6. Genetic loci; *H-2A, H-2E* clusters in mice; *HLA-DP, HLA-DO, HLA-DR* and several others in humans.

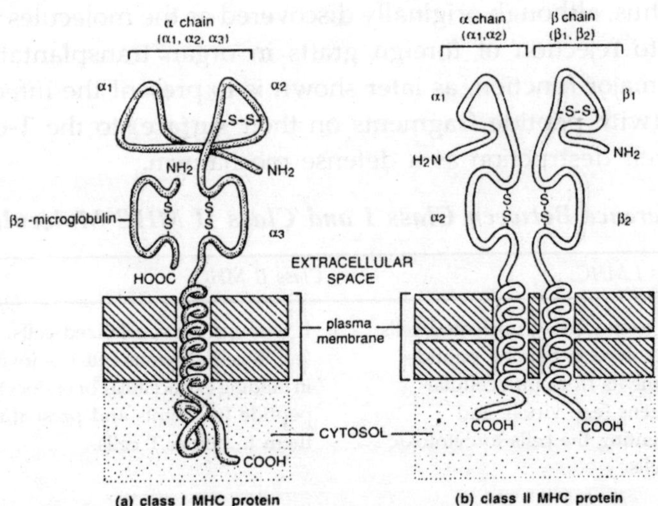

H-2 (mouse) and HLA (humans) gene complexes, showing the location of gene loci encoding the transmembrane units of class I and class II MHC proteins. There are three loci for class I proteins each in mouse (H2K, H-2D, H-2D) and in humans (HLA-A, HLA-B, HLA-Q, but the number of loci for class II differs in mouse (H2-A, H2-B) and humans (HLA-DP, HLA-DQ, LA-DR; only three are shown, but actually there are more than three) (redrawn from Alberts et al., "Molecular Biology of the Cell", 1994).

MHC Molecules : Kinds

There are two types of MHC molecules (i) class I MHC molecules and (ii) class II MHC molecules. Both are transmembrane heterodimers, whose extracellular-NH2 terminal domains bind to the peptide fragment of the antigen for presentation to the T-cells. Somen of the characteristic features of these two types of MHC molecules are summarized in Table . Each class I MHC gene encodes a single polypeptide called a having the following domains: (a) carboxy-terminal intracellular domain, (b) a transmembrane α -helix and (c) three extracellular globular domains $(\alpha_1,\alpha_2,\alpha_3)$. Each polypeptide is associated (non-covalently) with a β_2-microglobulin, which is extracellular and is coded by a gene lying outside the main *MHC* gene cluster (Figs.). Two domains of class I MHC i.e. β_2 and α_3 are homologous to an Ig-domain and α_1 and α_2 make a pocket or groove for binding of peptide fragment (8-10 amino acids long) of the antigen.

Class II MHC molecule is also a heterodimer having α and β polypeptides, each having a conserved Ig-like domain (α_1,β_2) close to the membrane and an amino-terminal domains (α_1,β_1) away from the membrane, providing site for the binding of antigen peptide fragment (Fig.). The presence of Ig-like domains suggest common evolutionary origin of MHC molecules, T-cell receptors and antibodies. The three-dimensional structures of class I and II MHC molecules have also been worked out. It was shown that the groove of class I MHC formed by α_1 and α_2 domains of α polypeptide (Fig.) accommodates 8-10 amino acids, while the groove of class II MHC molecule, formed of α_1 and β_1, accommodates 15-24 amino acids. The variety in both MHC molecules is such that they can bind to unlimited variety of peptides and present them to cytotoxic T-cells and helper T-cells.

Co-receptors (CD4 and CD8) of T-Cell : The T-cells have T-cell receptors and CD-complex to facilitate interaction between the T-cells and the target cells having peptide-MHC complex on their surface. This interaction is not strong enough and is unstable, so that accessory receptors called co-receptors are present to stabilize the interaction through cell-cell adhesion. These co-receptors also activate T-cells by generating their own intracellular signals and are also needed in T-cell development. They bind to non-variable part of specific MHC and are thus invariant. The two co-receptors are (i) CD4 on helper T-cells and binding class II MHC, and (ii) CD8 on cytotoxic T-cells, and binding class I MHC (Fig.). Both the co-receptors are single transmembrane proteins, their cytoplasmic tail being associated with members of Src family of tyrosine-specific protein kinases called the Lck protein. This kinase phosphorylates several cellular proteins thus activating the T-cell. CD4 also functions as a receptor for the HIV causing AIDS, thus permitting the virus to enter helper T-cells, paralyzing the immune system. Some of the accessory proteins (e.g. co-receptors) displayed on the surface of T-cells and the ligands displayed on the surface of target cells are summarized in Table.

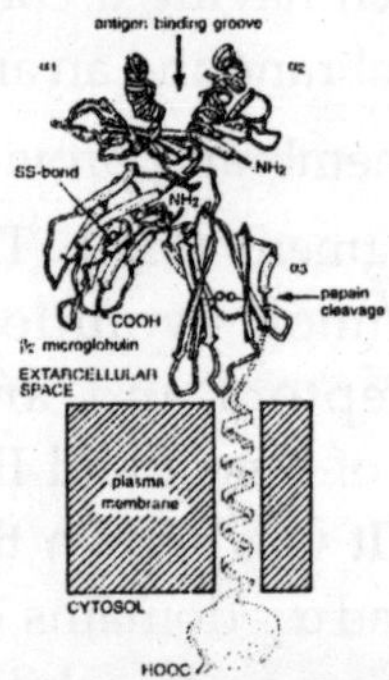

(a) structure of a human class I MHC protein showing peptide binding groove (side view)

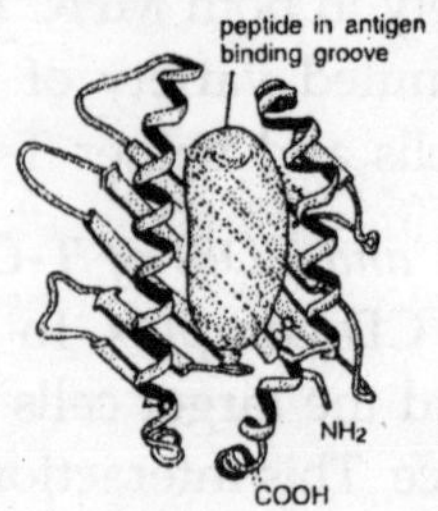

(b) peptide-binding groove (top view)

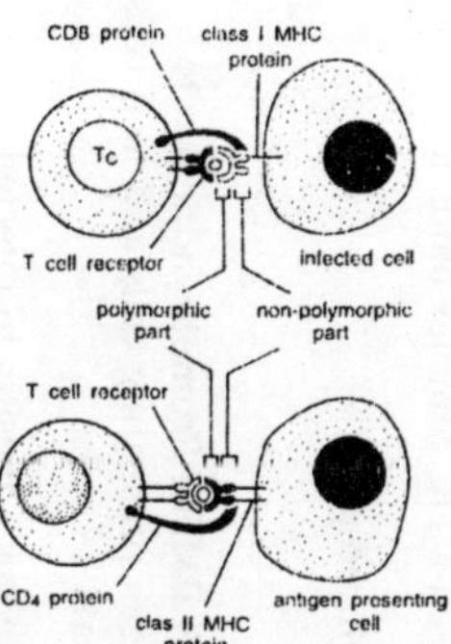

CD4 and CD8 co-receptors on the surface of T-cells; they bind to the non-variable region of the same MHC molecule, which is engaged by T-cell receptor, thus bringing them closer to T-cell receptors during T-cell activation (redrawn from Alberts et al., ' 'Molecular Biology of the Cell", 1994).

MHC restriction and Cytotoxic T-Cells : In 1974, for the first time, it was shown that MHC molecules present foreign antigens to cytotoxic T-cells, which kill the infected target cells having class I MHC, both in the mice and in culture. It was also shown that the T-cells recognize a specific antigen only when it is associated with a specific form of MHC molecule (Fig.). The phenomenon is known as MHC restriction and its importance was recognized in 1996 by the award of a Nobel prize in Medicine to Rolf Zinkernagel and Peter Doherty. It was later found that T-cells recognize the degraded fragments of internal proteins of a virus, and not the intact virus. This meant that only a small fraction of fragments need to reach the cell surface to cause an attack by a cytotoxic T-cell. It was also shown that the peptide fragments are transported from the cytosol to the ER-lumen, where they bind to MHC molecules (MHC molecules are made in ER lumen only) and are carried to the cell surface.

The viral antigens are degraded in the cytosol by proteasomes (enzyme complexes) containing two subunits encoded by genes within the MHC gene cluster. The peptide fragments are transported to ER lumen with the help of channels, that belong to the family of ABC transporters (ABC stands for ATP binding cassette; AB transporters are ATPases).

Surface of T-cells and Some Accessory Protiens
(modified from Alberts 'Molecular Biology of the Cell' 1994)

*Protein**	*Approximate molecular weight*	*Super family*	*Expressed on*	*Ligand on target cell*	*Functions*
CD3	γ chain = 25,000	Ig		—	Helps transduce signal when
	δ chain = 20,000	Ig	All T-cells		antigen-MHC complex binds to
	ε chain = 20,000	Ig			T-cell receptors
	ε chain = 16,000				
CD4	55,000	Ig	Helper T-cells		Class IIMHC Promotes adhesion to antigen presenting cells and to B cells; signals T- cells
CD8	70,000 (homodimer or heterodimer)	Ig	Cytotoxic T-cells	Class I MHC	Promotes adhesion to infected target cells; signals T-cells
CD28	80,000 (homodimer)	Ig	Many helper and cytotoxic T-cells	B7	Provides signal to some T-cells
LFA-1	α_1 chain = 190,000	intergrin	Most white blood		ICAM-1 Promotes cell-cell adhesion
	β_2 chain = 95,000		cells, including T-cells		

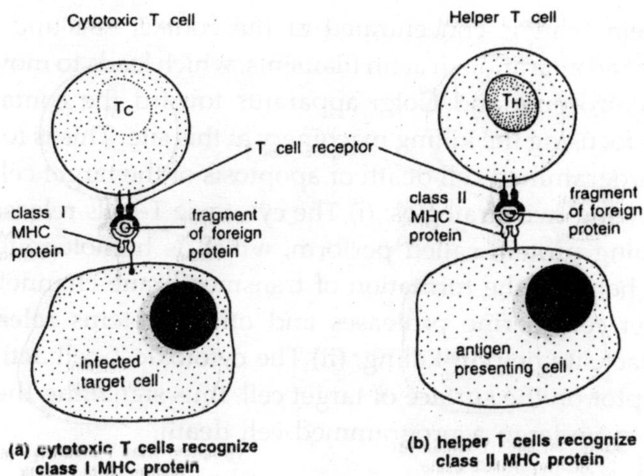

cytotoxic and helper T-cells with different MHC molecules (redrawn for mAlberts et al., "Molecular Biology of the Cell", 1994.)

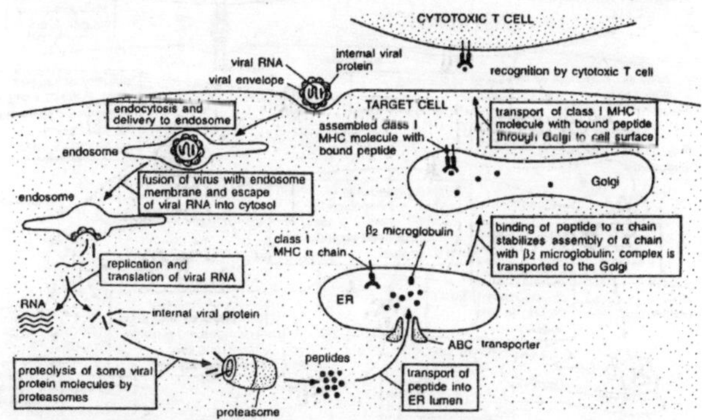

The processing of viral protein for presentation to cytotoxic T-cells

The details of the pathway of virus from outside to the host cell followed by presentation of viral peptides to the cell surface by MHC molecules is shown in Figure. Antiviral response of cytotoxic T-cells is enhanced when T-cells, activated by the antigen, secrete signaling molecules like γ-interferon, which puts into action the entire machinery needed for presentation of viral antigen to cytotoxic T-cells. The killing of the target cell by cytotoxic T-cells is facilitated by a protein called talin, that links T-cell receptors with cortical actin filaments. The protein talin is concentrated at the contact site and brings about an alteration in actin filaments, which leads to movement of centrosome and Golgi apparatus toward the contact site, thus focusing the killing machinery at this site. This is followed by programmed cell death or apoptosis of the target cell using following two strategies: (i) The cytotoxic T-cells release pore-forming protein called perform, which is homologous to C9 and helps in the formation of transmembrane channel in the target cell. Serine proteases and other proteins released at contact site help in killing, (ii) The cytotoxic T-cell activates a receptor on the surface of target cell, thus signalling the target cell to undergo a programmed cell death.

T-Cell Activation and T-Cells which Help

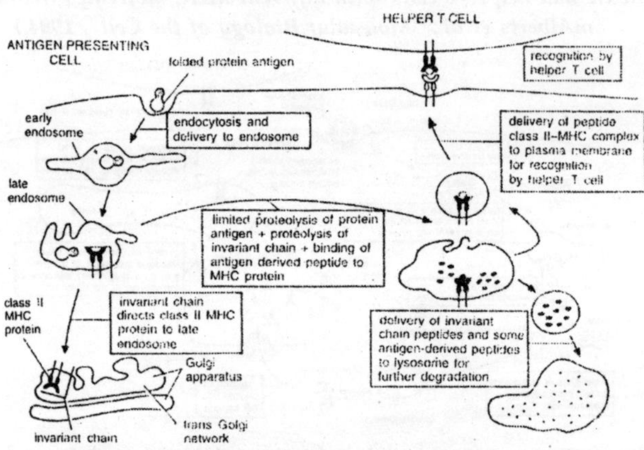

The processing of an extracellular protein antigen for presentation to a helper T-cell (redrawn from Albert et al., "Molecular Biology of the Cell", 1994).

Helper T-cells do not kill the infected target cells (as done in cytotoxic T-cells), but instead either stimulate macrophages to destroy pathogen or help other lymphocytes (e.g. B-cells) to respond (by antibody production in case of B cells). Acquired immuno-deficiency syndrome (AIDS) provides the most important example about the role of helper T-cells in immune system. The HIV (human immuno-deficiency virus) causes depletion of the number of helper T-cells by an unknown mechanism, thus destroying the immune system and rendering the patient susceptible to infection by micro-organisms (even though the virus is not really dangerous otherwise) eventually leading to the patient's death.

The helper T-cells, through their receptors, are recognized by the class II MHC associated with an antigen and displayed on the surface of a target cell (Fig.). This target cell earlier took part in the endocytosis of a protein/antigen. In other words the class II MHC molecules acquire peptide fragments from degraded endocytosed proteins and reach the surface of the target cell to present them to helper T-cells (Fig.). The helper T-cells need other additional signals for activation. These activated helper T-cells in their turn activate macrophages or lymphocytes (e.g. B cells).

Although class II MHC associates with degraded peptide fragments as in case of class I MHC (involved in presentation to cytotoxic T-cells), but there are following two major differences : (i) the peptides are derived from extracellular microbes or their products rather than synthesized inside the cell under the influence of a viral genome (as in case of class I MHC involved in cytotoxic T-cells); these microbes or their products are ingested by antigen-presenting cells and degraded in the acidic environment of endosomes; they never enter ER lumen, although class II MHC molecules are also synthesized in ER, just like class I MHC molecules.

The class II MHC molecule is actually pre-assembled and through trans-Glogi network reaches the late endosome, where it associates with peptide fragments, using its binding groove. The binding groove is initially occupied by a 'dummy' peptide called CLIP (derived from an invariant chain) thus preventing other ER peptides to bind. In the endosome, CLIP is released by proteolysis so that the groove is now accessible to peptide fragments derived from endocytosa proteins. Acidic conditions promote dissociation of CLIP and association of the foreign invaders in the groove of MHC class II molecules. At low pH (as found in the vesicle), foreign peptide loading is enhanced in the endosome-like compartment, where CLIP is replaced by this peptide.

Thus the class I MHC molecule presently peptides originating in cytosol, while class II MHC molecule presents peptides that come from extracellular space through endocytosis. A number of other proteins also take part in the activation of helper T-cells. Of these other cell surface molecules, B7 and ICAM expressed on target cell are important and play crucial role in activating T-cells (Table). Not all kind of cells can work as antigen-presenting cells. Only the following types of cells can take part in antigen presentation : (i) interdigitating dendritic cells in lymphoid organs; (ii) langerhans cells ill skin; (iii) B-cells and macrophages; (iv) thymol epithelial cells.

Activation of helper T-cell (e.g. interleukin-1 or IL -1 and B7+CD28) and the Need of Two Signals : As discussed earlier, a complex is formed at the site of contact of helper T-cell with the targets cell. This complex consisting of T-cell receptor, co-receptors, CD3 complex and tyrosine kinase provides one of the two signals for activation of T-cells.

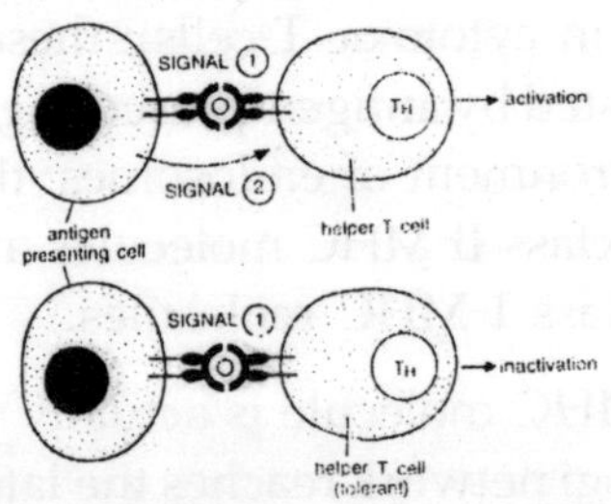

Two signals are required for helper T-cell activation (redrawn from Alberts et al., "Molecular Biology of the Cell", 1994).

The second signal is provided by either of the following two pathways : (i) a chemical signal (e.g. interleukin-1 or IL-1) is secreted by the antigen-presenting cell or (ii) a protein B7 is bound to the plasma membrane of these antigen-presenting cells. B7 is recognized by a co-receptor called CD28 (a member of Ig superfamily) present on the surface of helper T-cell. If the above two signals are available, the helper T-cells are activated, and secrete a variety of interleukins (except IL-1). It has also been shown that in the absence of second signal, after receiving the first signal, the helper T-cells become transformed in such a way, that they can not be activated, even if both the signals become available later on (Fig.). In other words T-cells become tolerant.

Helper T-cells and their Proliferation Due to Interleukins (Lymphokines or Cytokines) : Interleukins are peptides or proteins which are secreted by T-cells and other cells. They mediate local interactions between white blood cells (leucocytes), but do not bind antigens. Interleukins secreted by lymophocytes are also called lymphokines. Interleukins are also made by non-blood cells (e.g. IL-1 and IL-6) and act on many types of target cells. In view of this, they are more accurately called cytokines. Some of these interleukins and their sources, targets and actions are summarized in Table. After having received the twin signals as above, the helper T-cells stimulate their own proliferation. For doing this, they simultaneously secrete (i) interleukin-2 or IL-2 (a growth factor) and (ii) cell surface receptors that bind IL-2. The binding of *IM-2* to their receptors stimulates T-cells to proliferate, after the T-cells dissociate from the antigen- presenting target cells (Fig.). In fact IL-2 can bind and stimulate proliferation of any neighbouring cells (including cytotoxic T-cells) that possess IL-2 receptors on their surface, but the IL-2 receptors are synthesized only due to induction by antigens on the target cell surface. The above principle has actually been used for T-cell hybridomas for research and other purposes (e.g. isolation and study of T-cell receptors and their genes).

Besides their own proliferation, helper T-cells also activate cytotoxic T-cells and B cells. This is achieved by secretion of interleukins, which differ for two different kinds of helper T-cells. Thl cells secreting IL-2 and y interferon (IFN-y) and Th2 cells secreting IL-4 and IL-5. While Thl cells help cytotoxic T-cells, Th2 cells help B-cells and eosinophils.

Some Interleukins and their Characteristics (modified from Alberts 'Molecular Biology of the Cell', 1994)

Interle-ukln (IL)	*Approximate molecular weight*	*Source*	*Target*	*Action*
IL-1	15,000	Antigen-presenting cells	Helper T-cells	Helps activate
IL-2	15,000	Some helper T-cells	All activated T-cells and B-cells	Stimulates proliferation
IL-3	25,000	Some helper T-cells	Various hemopoietic cells	Stimulates proliferation
IL-4	20,00	Some helper T-cells	B-cells	Stimulates proliferation, maturation, and class switching to IgE and IgGl
IL-5	20,000	Same helper T-cells that make IL-4	B-cells, eosinophils	Promotes proliferation and maturation
1L-6	25,000	Some helper T-cells and macrophages	Activated B-cells, T-cells	Promotes B-cell maturation to Ig secreting cells; helps activate T-cells
γ-Interferon	25,000 (dimer)	Same helper T-cells that make IL-2	B-cells, macrophages, endothelial cells	Activates various MHC genes and macrophages

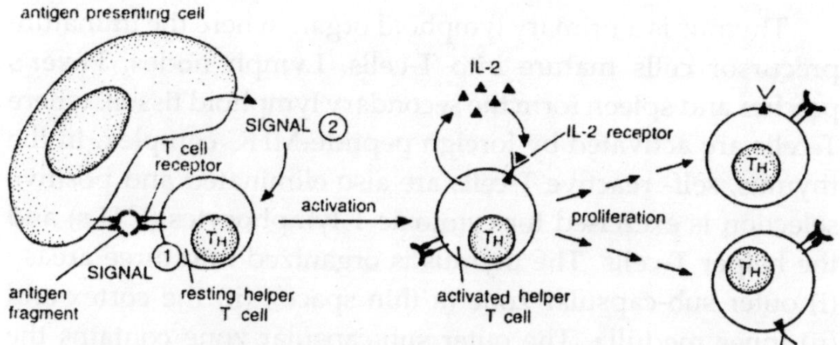

The stimulation of T-cell proliferation by interleukin-2 (IL-2), which binds to its receptor; both IL-2 and its receptor are secreted by the T-cells themselves, when they receive the two signals (redrawn from Alberts et ai, "Molcculur Biology of the Cell", 1994).

Maturation/selection of Developing T-cells and MHC Restriction : MHC restriction means that the T-cells recognize foreign antigens in association with only the self MHC molecules and not in association with foreign MHC molecules. In transplantation, the graft rejection is actually due to response of T-cells against foreign MHC molecule associated with self peptides, which mimics self MHC complexes with foreign antigen. This restriction is due to a positive selection of T-cells during their development in the thymus. During this selection, only those T-cells survive, which recognize peptides presented by self MHC; other T-cells die. MHC restriction is thus a property acquired during development of T- cells in the thymus.

A negative selection on developing T-cells is also exercised for self tolerance, i.e. against recognition of self peptides, because they are supposed to recognize only the foreign peptides associated with self MHC. If these T-cells start recognizing self peptides with self-MHC molecules i.e. if they lack self tolerance, it will be disastrous for any individual. These phenomena of positive and negative selection of T-cells during their development in the thymus are shown in Figure.

Thymus is a primary lymphoid organ, where the immature precursor cells mature into T-cells. Lymph nodes, Peyer's patches and spleen form the secondary lymphoid tissue, where T-cells are activated by foreign peptide-MHC complex. In the thymus, self- reactive T-cells are also eliminated and positive selection is exercised for cytotoxic T lymphocytes (CTLs) and the helper T-cells. The thymus is organized into three areas : (i) outer sub-capsular zone (a thin space); (ii) the cortex and (ii) inner medulla. The outer sub-capsular zone contains the most immature $CD4^-$, CD8-T-cell precursors received from the bone marrow. These $CD4^-$ $CD8^{-\prime}$ precursors enter the cortex, where the epithelial cells express class I and class II major histocompatibility complex (MHC) encoded molecules. In the cortex, the T-cell precursors also express CD4 and CD8 and allow rearrangement of TCR (T-cell receptor) genes and B genes, so that each cell will have a different rearranged TCR-B heterodimer.

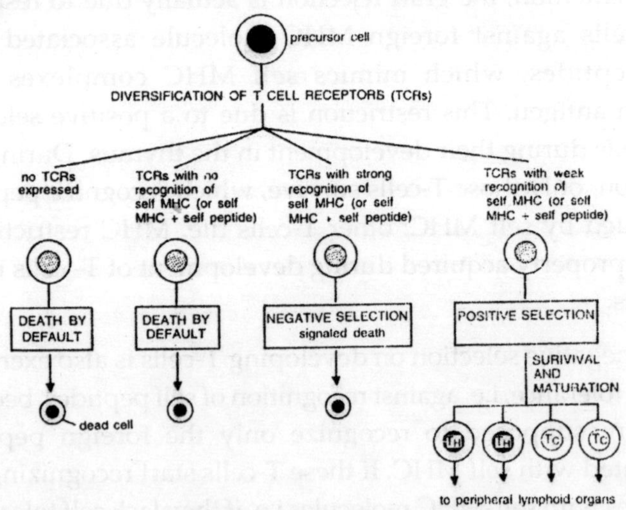

Positive and negative selection of T-cells in thethymus (see text for details; redrawn from Alberts et al., "Molecular Biology of the Cell", 1994).

These precusor T-cells during maturation undergo positive and negative selection, so that the thymocytes ($CD4^+$ $CD8^+$)

with TCR having low avidity for self peptide-MHC complex will be positively selected, (those with high avidity will be eliminated here or at a later stage; thymocytes with TCR having no avidity for self peptide-MHC complexes will also die through apoptosis). These selected cells will now differentiate into $CD4^+$ $CD8^-$ (helper T-cells), if its TCR is specific for a self peptide-class II MHC complex or will differentiate into *CD4*$^-$ $CD8^+$ (CTLs), if its TCR is specific for a self peptide-class I MHC complex. These selected T-cells will exit the thymus and enter the secondary lymphoid tissue (lymph nodes, spleen, etc.). The selected T-cells ($CD4^+$ $CD8^-$ and $CD4^-$ $CD8^+$) enter the inner medulla of thymus. The T-cells that are autoreactive having strong avidity for self peptide-MHC complexes may still be present in the population of T-cells that were positively selected in the cortex. These are still $CD4^+$ $CD8^+$ and undergo elimination either in the cortex or in the medulla before differentiating into $CD4^+$ $CD8^-$ or $CD4^-$ $CD8^+$ types. From the medulla these cells exit the thymus and enter the blood reaching the secondary lymphoid tissue (lymph node, spleen, etc.), where they will be available to be activated by a foreign peptide-MHC complex.

Paradox in Peripheral Tolerance

Since T-cells, that are self reactive, are eliminated in the thymus, one would expect that self-reactive T-cells will not be found in the population of peripheral T-cells. However, the T-cells do not come across all the possible antigens within the thymus. Therefore, self antigens that are not expressed in the thymus should express elsewhere in the body areas patrolled by mature T-cells. Tolerance to these antigens is difficult, because the peripheral T-cells are programmed to respond to these antigens, without making any distinction between self-antigens and foreign antigens. How to silence the peripheral T-cells that are specific for extra-thymic self antigens, without being able to eliminate them? This is achieved by stopping the T-cells and B-cells in the secondary lymphoid tissue, while they circulate between the blood and the lymph. Here the

T-cells are inactivated by a process called clonal anergy (see later for clonal anergy).

B-cells, MHC Molecules and Immune Response (Ir) Genes : The cell-dependent immune response to an antigen also depends on a single gene called immune response *(Ir)* gene, and the responses to different antigens are controlled by different *Ir* genes. It was later shown that these *Ir* genes are actually MHC genes and that the lack of response earlier attributed to mutation in *Ir* gene was actually due to the absence of corresponding MHC molecule or due to assembly of self MHC molecule-foreign peptide combination that sometimes transform and mimic-self MHC molecule—self peptide combination.

CD4⁺ Helper T lymphocytes (Thl and Th2) and Two Subsets : Helper T cells (also called $CD4^+$ cells) are involved in two types of immune responses : (i) strong delayed-type hypersensitivity (DTH) associated with low level of antibody (e.g., tuberculoid form of leprosy) and, (ii) weak DTH associated with high level of antibody (e.g., lepromatous leprosy). The above two responses can be related with two types of mouse $CD4^+$ T^- cell clones that were recognized for the first time in 1986, on the basis of their pattern of cytokine production.

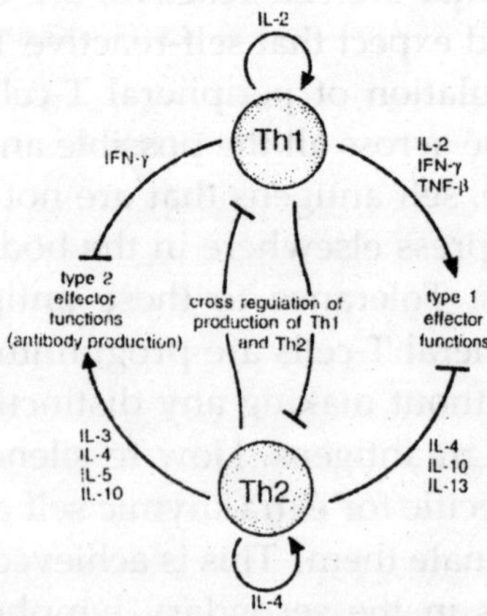

Two types of helper T-cells, Th1 and Th2, showing the molecules secreted by them, their effector I functions and other interactions.

These two types of helper T-cells also differ in their effector functions (see later). These were called Thl and Th2. Of these two types, Thl clones produce interleukin (IL)-2, interferon (IFN)- γ and tumour necrosis factor (TNF)- β. Similarly, Th2 clones produce IL-3 IL-4, IL-5, IL-10 and IL-13 (Fig.). IL-10 was originally described as the product of Th2 but is now known to be produced by both (Thl and Th2). Later work, however, demonstrated that T-cells can not be easily classified on the basis of the above criteria, since each may exhibit complex pattern of cytokine production. T-cells- producing various mixtures of cytokines have been called 'ThO (Th zero). This heterogeneity is still greater in human T lymphocytes than in the murine cells. In view of this, some immunologists prefer to call the two subsets as Thl like' and 'Th2 like' or 'Thl-dominant and 'Th2-dominant'. Despite the difficulty in classification, Thl and Th2 designations are being used and perhaps serve a useful purpose.

It is also known that Thl cells produce strong DTH reaction, while Th2 cells produce weak DTH reaction involving high level of antibody production. It was also shown that each type produces cytokines for its own proliferation (autocrine growth factor IL-2 for I Thl and IL-4 for Th2) and differentiation. They also produce cytokines, that cross-regulate each other's development and activity (IFN-γ suppressing Th2 and IL-4 and IL-10 suppressing Thl). The net result is that once a T-cell immune response begins to develop along one pathway, it gets progressively polarized in that direction only, due to autoamplification and cross-regulation (Fig.). Ever since, these Thl and Th2 types were recognized for the first time in 1986, considerable work has been done and concepts developed. These concepts can now be used for developing prophylactic vaccines and therapeutic approaches for many immunological diseases. For instance, leishmaniasis remains a paradigm for functional dichotomy of helper T-cells, Thl associated with resistance and Th2 with susceptibility to this disease.

Thl and Th2 effector functions. As indicated above, Thl cells are primarily responsible for cell-mediated immunity or strong DTH (killing through phagocytosing the antigen presenting cells, e.g. cells containing viruses) and Th2 are involved in humoral immunity or weak DTH (they help B cells and other cells in producing antibodies). However, IFN-y secreted by Thl cells induces the switching of B lymphocytes to several IgG isotypes, thus questioning the above classification based on their above effector functions. Therefore, their effector functions need to be classified on the basis of biological activities of cytokines produced by them.

The principal cytokine of Thl is IFN-y, which activates macrophages for microbicidal action and also stimulates production of IgG antibodies. Th-2 cells, on the other hand, produce IL-4 and IL-5; IL-4 induces B-cell switching to Ig-E production and IL-5 activates eosinophils. IgE and eosinophil-mediated defense is important for eliminating some, but not all helminths. Th2 cells also help B lymphocytes and stimulate the production of IgM and IgG isotypes. The Th2 cells also inhibit inflammatory responses (Fig.).

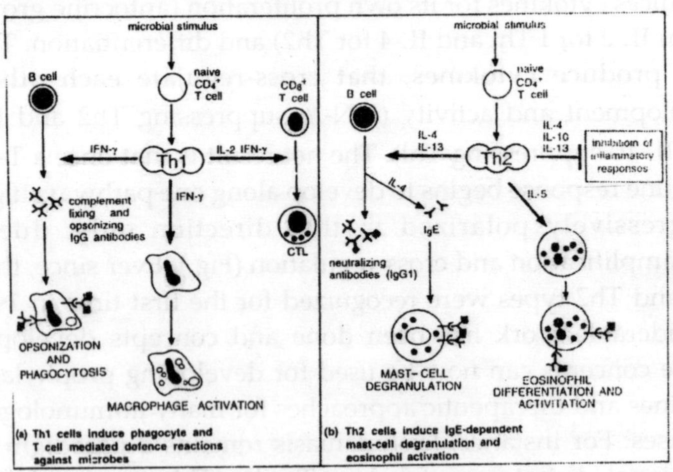

Different steps involved in microbial-stimulated reactions of Thl and Th2 helper T-cells.

CD4⁺ Helper T-cells (Th1 and Th2) Diseases and their Subsets : The Thl/Th2 dichotomy has been found to be associated with many diseases. This is due to the critical role played by individual cytokines in controlling pathological immune responses. These diseases include infectious diseases like leischmaniasis, allergic diseases (involving IgE and mast cells) and autoimmune diseases. Evidence is available suggesting that modulation of Thl/Th2 balance by administration of recombinant cytokines or cytokine antagonists alters the outcome of the disease. For instance, IL-12 if administered at the time of infection enhances resistance to many intracellular pathogens (protozoa, bacteria, fungi and some viruses). Similarly, when IL-12 is used as a vaccine adjuvant with sensitizing doses of antigen, it converts Th2 pattern to Thl pattern, thus promoting resistance to intracellular infection, and suppressing Th2 dependent pathology. IL-12 can also be used for treatment of allergy, cancer, etc., and is being tested in anticancer vaccine protocols. IL-10 has also been used to suppress lipopolysaccharide-induced endotoxaemia and inflammatory bowel disease in experimental animals. Selective induction of Th2 cells has been proposed for treatment of tissue autoimmune diseases.

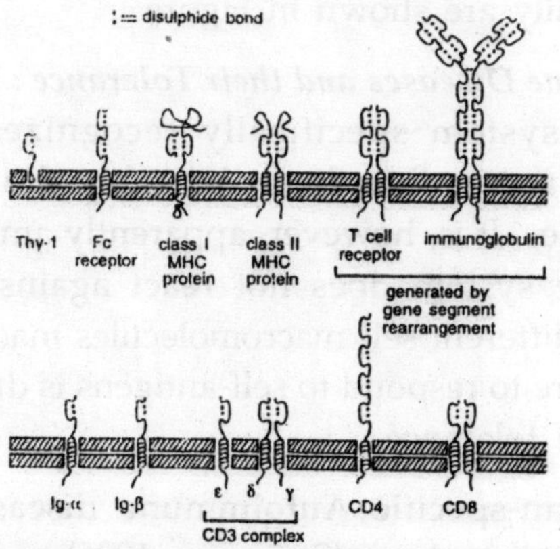

Some of the member proteins belonging to the Ig superfamily (redrawn from Alberts et al., "Molecular Biology of the Cell', 1994).

As an alternative to administration of cytokine or cytokine antagonists, Thl/Th2 balance may also be altered by specific treatment. For instance, oral administration of antigens favours the development of T-cells that produce TGF- P (transforming growth factor) and Th2 cytokines. This is the basis for 'oral tolerance' protocols for preventing autoimmune diseases. In future, better understanding of Thl/Th2 dechotomy will help in the development of prophylactic vaccines and therapy for many diseases associated with 111 dichotomy.

Members of Ig Super-family

The proteins generally involved in cell-cell recognition and those involved in antigen recognition in the immune system contain elements related to each other. These elements belong to an Ig superfamily which includes till following classes of proteins : (i) antibodies! (ii) T-cell receptors; (iii) MHC proteins; (iv) cm receptors (CD4, CD8, CD28); (v) invarial polypeptide chains of B and T-cell receptoul (vi) Fc receptors on lymphocytes and other WBCs. Each Ig-like domain has 70-110 amioil acids, and is a folded structure made up of two antiparallel β sheets, that are stabilized by disulphide bonds. Some of these proteins of lg superfamily are shown in Figure.

Autoimmune Diseases and their Tolerance : We know that the immune system specifically recognizes and reacts aggressively to a virtually unlimited number of different foreign macro molecules. It is, however, apparently amazing that the same immune system does not react against the tens of thousands of different self macromolecules made by the host cells. The failure to respond to self-antigens is due to acquired immunological tolerance.

Many organ-specific Autoimmune diseases involving supramolecular complexes (Steinman, 1996) and some organs relating to them

	Disease	*Supramolecular complex*
1.	Myasthenia gravis	Acetylcholine receptor : all subunits
2.	Thyroiditis	Thyroid-stimulating hormone receptor, thyroglobulin, thyro-peroxidase
3.	MS	Myelin : MBP proteolipid protein, myelin oligodendroglial protein, aB crystallin
4.	Insulin-dependent	Islet antigens : GAD, heat shock protein
	diabetes mellitus (IDDM)	65
5.	Primary biliary cirrhosis	Mitochondrial branched chain keto acid dehydrogenase complexes
6.	Uveitis	Photoreceptor-interphotoreceptor retinoid-binding protein, S-antigen of rout outer segment

One would like to understand how does the immune system distinguishes foreign molecule from self molecules and then react only against such foreign molecules. There are two possibilities : (i) one possibility is that all animals may inherit genes that encode receptors for foreign antigens and not for self B antigens; (ii) the second possibility is that the immune system of an animal is inherently capable of responding to both foreign and self antigens, but during the development it learns to respond to foreign antigens only; this means that the differential response to foreign antigens is as acquired property. Research conducted during the last few decades has conclusively shown that the second alternative is correct and that the immune system acquires the attribute of responding to only the foreign antigens. This was initially demonstrated by the following observations : (i) When tissues from one individual are transplanted into the body of another, they are recognized as foreign and therefore destroyed; (ii) In dizygotic twins (genetically different), if cells of one mouse are introduced into the body of another (at neonatal stage, before immune system matures), then no response to these foreign cells was observed for a long time.

It was also shown that the continuous presence of self antigen is necessary for the acquired tolerance. For instance, in mouse, the immune response does not react against complement protein C5, but in a mutant lacking C5, if this protein is introduced from outside, the immune system responds. Similarly if C5 is removed, the animal regains the ability to respond to it within weeks or months. It means that the system is inherently or genetically capable to react to self antigens, but learns not to do so. This acquired tolerance can be achieved in either of the following two ways : (i) killing of self-reactive lymphocytes (clonal deletion); (ii) functionally inactivating the self reactive lymphocytes, but leaving them alive (clonal anergy).

Tolerance of self antigens, however, sometimes breaks down, so that T-cells or B-cells or both start reacting to their own antigens. This causes diseases described as autoimmune diseases (Table 24.5). It is believed that one of the reasons for autoimmune disease is the failure to eliminate self-reactive lymphocytes during their development. Second reason is the creation of physical or functional holes in the B and T-lymphocyte receptor repertoires. For instance, diabetes (insulin-dependent diabetes mellitus or IDDM) is a chronic disorder, which results from destruction of insulin-producing P cells of pancreas islets. Initially, T-cells and other inflammatory cells invade the islets, eventually destroying them. It was possible to isolate a clone of $CD4^+$ T-cells, which alone are capable of conferring disease. It is these cells perhaps, which escape

elimination during T-cell development and thus cause the disease. The T-cell receptor (TCR) on these T-cells recognize autoantigen. Transgenic mices have been produced, expressing autoreactive TCR and were designated BDC2.5. In these transgenic mice, there was no negative selection exercised during the development of T-cells in the thymus, and the pathogenesis followed the following course : (i) In the first three weeks, there was no manifestation of the disease, and pancreatic islets were completely free of infiltration of the reactive T-cells. (ii) After 3 weeks, insulitis sets in abruptly and involves all islets. (iii) Diabetes appeared only later and widely spread over 10-25 weeks, a pattern also observed in non-transgenic diabetic mice.

The above experiments suggested that the disease does not really result so much from the failure to eliminate self reactive T-cells, but from the breakdown of secondary systems that keep these self-reactive cells in check. It means that the self-reactive cells and cells with anti-self specificities do occur even in disease- free individuals, but they ignore target tissues. This is achieved through following restraints that these self-reactive cells observe. (i) Their ability to respond to antigen is diminished by mechanisms that tune down signaling of their antigen receptors, (ii) Their ability to carry out inflammatory effector functions can be inhibited, (iii) Their ability to migrate and persist is constrained.

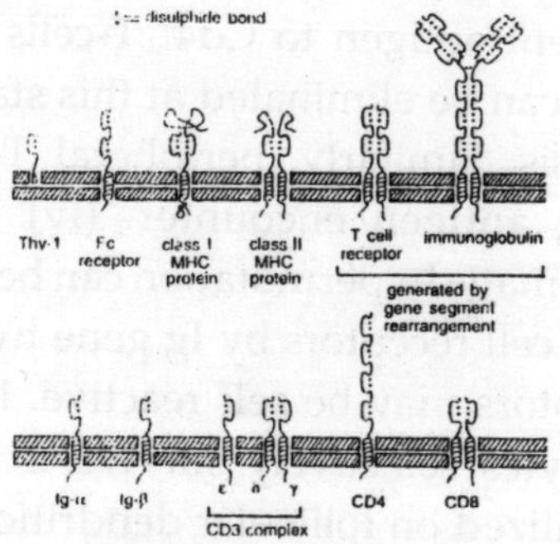

Balance between

Autoimmunity, Tolerance and their Balance : A balance between the tolerance and immunity is actually maintained in an immune system. This involves positive and negative selections, the former involving survival, activation and expansion of one class of T-cells, and the latter involving inhibition and elimination of another class of T-cells. A shift in this balance on either side leads to resistance or susceptibility to an autoimmune disease.

Clone Elimination and Tolerance

The elimination of self-reactive clones of B and T-cells can be exercised at different stages and may include the following : (i) Elimination within bone marrow and thymus has been demonstrated both for B-cells and T-cells. It has been shown in many cases that certain clones of immature B and T-cells are physically eliminated. They do so by programmed cell death or apoptosis. (ii) Elimination after arrival in peripheral lymphoid tissues is also exercised in many cases. This was illustrated by using transgenic animals, where B-cells recognized self antigen (transgene encoded), the serum lysozyme. These B-cells were eliminated by apoptosis, 1-2 days after migrating to spleen and lymph node. However, a much smaller fraction of newly formed T-cells is exported from the thymus to the peripheral lymphoid tissue and eliminated there, (iii) Elimination after activation in peripheral lymphoid tissues can also be achieved. For instance, B-cells that present antigen to $Cd4^+$ T-cells in the peripheral lymphoid tissue can be eliminated at this stage through CD95 induced apoptosis. Similarly, peripheral T-cells can also be killed following antigen encounter. (iv) Elimination after activation and somatic hypermutation can be achieved through remodeling of B-cell receptors by Ig gene hypermutation. The remodeled receptors may be self reactive. It has been shown that the centrocytes selectively survive if they bind foreign antigens immobilized on follicular dendritic cells. However, if these centrocytes bind free antigens, they undergo death or apoptosis. The complement C3d components on the surface of dendritic cells play a role in this selection.

Clone Retuning and Tolerance

Self tolerance can also be achieved by altering self-reactive cells, instead of physically eliminating them. This is called clonal anergy. For instance, it has been observed that if B-cells are first exposed to a weak stimulus by an antigen lacking a co-stimulator, it becomes difficult to activate them by a subsequent immunogenic antigen challenge (e.g. exposure to anti-Ig antibody, serum lysozyme, single-stranded DNA). This anergy in B-cells reflects retuning of the antigen receptors on the surface of B-cells. These receptors in case of IgM are decreased by a factor of 20-50 due to selective block in their transport from ER.

In other cases, the receptors may be incapable of showing the normal response (tuning down the response). A similar phenomenon is also observed in T-cells. For instance, they become much more difficult to activate into clonal expansion, if presented with super-antigens, allo-antigens, peptide-antigens, in a weak, chronic or co-stimulator-deficient manner. Tuning down may involve down-regulation of the number of antigen receptor and co-receptor molecules on the cell surface. However, the anergic B-cells and T-cells are not inert, and may participate in immune response under certain conditions.

Inhibiting Clone Effector Functions and Tolerance : Sometimes, anergic self reactive B-cells may become aggressive and may tend to undergo clonal expansion by an antigen, due to the presence of non-specific mitogens such as bacterial lipopolysaccharide. This restoration of self reaction may be inhibited by continued binding of autoantigen to surface Ig receptors. B-cells exposed to anti-Ig antibodies may also exhibit a similar phenomenon.

Clone effector function may also be inhibited by the

inhibitory effect of Th2 on the development of IFN-y produced by Thl. This IFN-y promotes macrophages and also promotes B-cell isotype switching to IgG2a which also promotes inflammation.

Balance Between Tolerance and Immunity and Modecules Affecting it : Immune response is initiated by antigen presentation followed by the triggering of signal transduction pathway. The execution of this pathway needs effector molecules, which if defective cause a shift in the balance between tolerance and immunity. The autoantibody production due to lack of CD95-Fas or Fas ligand (required for elimination of self reactive clones) are examples of inherited defects causing immune diseases. Autoantibody production also occurs in children deficient in CD49L. Deficiency in IL-2 or IL-2 receptor also causes autoantibody production.

Bcl-2 and Bcl-x (involved in cell death or apoptosis) are other effector molecules that promote survival of clones reactive to foreign antigens. Deficiency of Bcl-2 causes B-cell lymphopenia and its overexpression leads to autoantibody production (due to inhibition of the elimination of self-reactive B-cells).

8

The Antibodies

Immunology deals with the defense mechanisms of animals against parasites, so that these animals are normally capable of resisting the infection by most pathogens. However, the biochemical and molecular basis of such resistance has been understood only during the last few decades. The pathogens are always recognized by the host cells through chemical interactions; a specific antigen or immunogen, present in the parasite's body, elicits in the host the production of a specific glycoprotein complex (an anitbody) or an immunoglobulin (Ig). The antibody or immunoglobulin is always a protein, covalently bound with an oligosaccharide.

In contrast, although most often, antigen is a protein, it may also be a polysaccharide or nucleic acid or any other substance. It is also possible that a foreign protein (not necessarily belonging to a pathogen) may act as an antigen so that when injected, it may induce antibody formation. Certain low molecular weight molecules called haptens, may though bind to antibodies, do not individually stimulate antibody production. However, they may become antigenic and stimulate antibody production, if they are tightly bound to certain macromolecules such as proteins, polysaccharides and nucleic acids. It should also be recognized that the entire surface of an antigen molecule is not necessary for its antigenicity; instead specific group of atoms called antigenic determinant or epitope, consisting of 5-8 amino acids, is needed for immune response and antibody production.

Natural Immunity and Acquired Immunity

'Natural Immunity' : 'Natural immunity' of an animal is also known as innate immunity, native immunity or inherited immunity. This relates to a general or non-specific type of resistance, which prevents, infection by different kinds of pathogens. The extent of this natural immunity differs in different organisms. For example, man can easily i get mumps while cats and dogs are immune to this disease. This level of natural immunity may vary not only between species but also between races or strains, and sexes. This may also be controlled by nutrition, hormones and many other factors.

'Acquired Immunity' : 'Acquired immunity' develops during the life time of an individual and refers to the immunity, which a specific individual displays against a specific pathogen. This is frequently related to the presence of antibodies or interferons in the blood. The most efficient acquired immunity is generally based on antibodies and may be either (i) actively acquired or (ii) passively acquired.

Actively acquired immunity. 'Actively acquired immunity' may be either natural or artificial. Actively acquired natural immunity results due to any infection from which a person recovers. During the infection, antibody production for that specific pathogen is stimulated so that when there is a subsequent infection by either the same or antigenically related pathogen, the antibodies assist in the body's defense.

Actively acquired artificial immunity is the most common

method of immunization or vaccination. The immunogens are injected in the body in controlled quantity to stimulate the production of immunoglobulins. Killed and attenuated strains of bacteria and viruses are now used widely for immunization against many diseases (e.g., typhoid, small pox, poliomyelitis, yellow fever, measles, etc.). Attenuated organisms produce mild infection and induce natural immunity. Vaccines are also being developed now by a variety of other methods involving recombinant DNA technology.

Passively Acquired Immunity : 'Passive Immunity' may also be acquired either by natural or artificial means. While passive immunity acquired by natural means involves the transfer of antibodies from mother to her unborn child, through the placenta during the later part of pregnancy, passive immunity of the artificial type refers to the original production of antibodies in some other individual (human or lower mammal) followed by injection of these antibodies in humans with the help of a needle or syringe. Several drug manufacturing companies are involved in the large- scale production of antibodies in horse and cows by active immunization to be used for passively acquired immunity in humans.

Immune Responses

The injection of an antigen into an animal may initiate several important changes in the animal. The first response is to eliminate the antigen through extensive phagocytosis in the liver, lung and spleen, so that nearly 90% of the antigen is removed from the circulation in about 10 to 20 minutes of its introduction. In the second phase, which lasts for about 4 to 7 days, the antigen is eliminated by enzymic hydrolysis and digestion. However, the most important phase in antigen elimination is the immune response involving formation of antibodies, which provide resistance to the animal body against infection.

System of Lymphoid (B cells and T cells) : The immune system of the body is monitored with the help of white blood cells called lymphocytes, which circulate between blood and the lymph system as illustrated in Figure (lymph is a colourless fluid and is circulated like blood with the help of vessels that make a lymphatic system just like blood circulation system).

The lymphocytes are of two types : B cells and T cells. The B cells synthesize antibodies causing agglutination of unwanted antigens, while T cells carry out cellular immune responses, where cells having unwanted antigens on their surface are attacked without any mediation by antibodies. The former response is called humoral immunity and the latter the cellular immunity (see Fig. for details).

The T cells interact with antigen through cell surfade proteins called T cell receptors, which are specially attached to the cell membrane. B cells and T cells both originate in bone marrow, which is the primary source of all lymphocytes. In mammals, B cells mature within bone marrow, but T cells migrate out and mature in special organs like thymus.

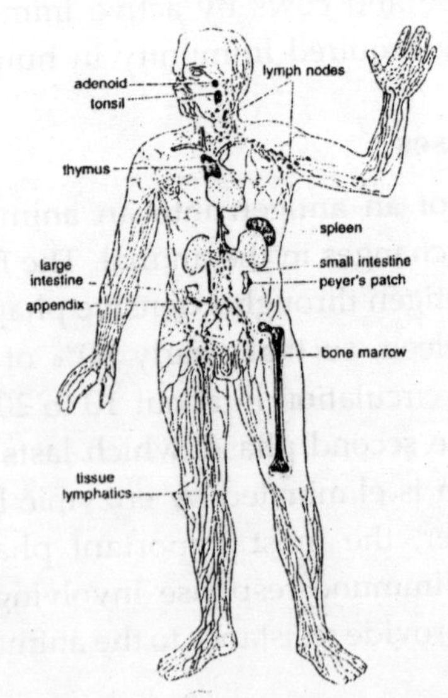

An outline of lymphoid system in a human being.

Once they mature, B cells leave bone marrow and T cells leave the organ like thymus and travel to secondary lymphoid organs like lymph nodes, tonsils and spleen. In these organs they can be activated to produce antibodies. These cells then pass to blood system and back again to the lymph.

While B cells can be stimulated by the free antigens in the circulation system of the body, T cells respond to only those antigens which are bound to cell surface, and that too only in the presence of another determinant protein called major histocompatibility complex or MHC. This helps the T cells in recognizing only the virus-infected cells of the host to which the T cells belong, and does not allow the recognition of the cells which are either normal uninfected cells of the same organism or the infected cells of another organism to which the T cells do not belong. This dual specificity of T cells is dependent on a large variety of T cell receptors available on T cell surface. Since T cell receptors resemble antibodies in structure and variability, their genes though distinct are similar to the Ig genes meant for the synthesis of immunoglobulins or antibodies.

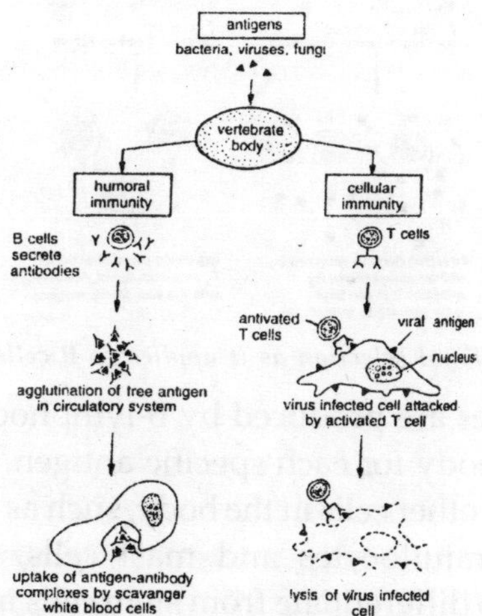

Immune response to an antigen involving humoral immunity through B cells and cellular immunity by T cells.

Theory of Clonal Selection

Various theories have been proposed regarding the formation of antibodies, although only the clonal selection theory seems to be acceptable in most situations. According to this theory, lymphocyte cells are able to make a specified antibody against a specific antigen. The cell cannot be making the antibody continuously but may get transformed into an antibody synthesizing cell when the antigen comes in contact of the proper lymphocyte cell. B cells, while not engaged in active synthesis of antibodies, carry specific antibodies on their surfaces. Due to these surface-bound antibodies, a specific antigen may get bound with specific B cells and may thus trigger the proliferation and maturation of that particular B cell. The cell responds with repeated cell divisions until a clone of antibody synthesizing cells called plasma cells is formed. These cells then form large number of same antibodies (Fig.), at a rate of 10,000 molecules per second per cell.

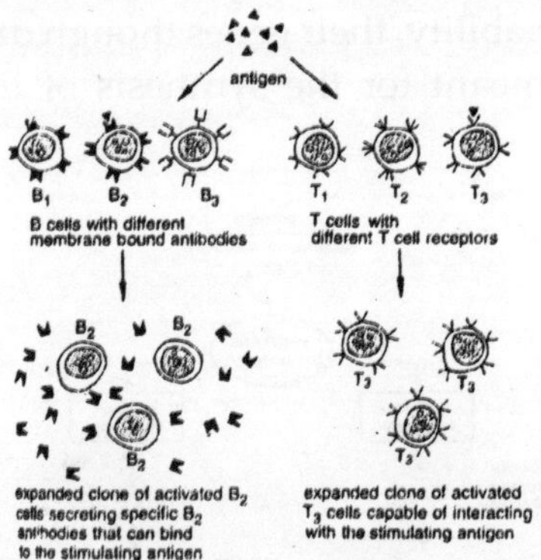

The theory of clonal selection as it applies to B cells and T cells.

The antibodies are produced by B-lymphocytes, and there is a specific antibody for each specific antigen. Antibodies are also produced by other cells in the body, such as 1-lymphocytes, macrophages, granulocytes and mast cells, which, like B-lymphocytes also differentiate from stem cells in bone marrow. A concerted effort of all these cells is required to produce the immune response.

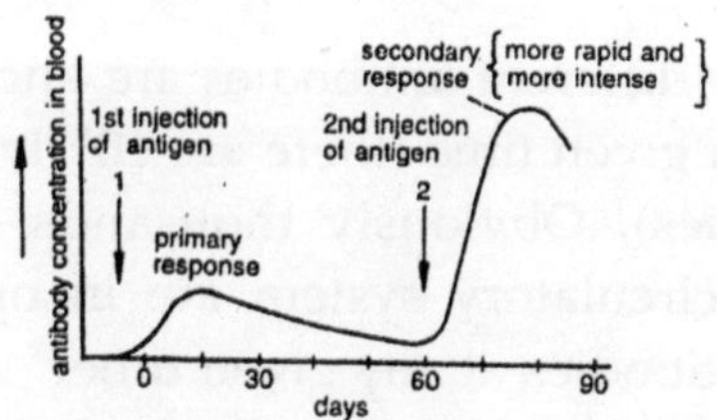

A graph showing the primary and secondary responses of antibody production in a rabbit.

When an animal first encounters an antigen, it produces a low primary response, since only few B cells or T cells can respond to antigen. However, the next encounter with the same antigen leads to a far more rapid and intense response called secondary response, since many cells are now available for response (Fig.). Due to the first exposure to antigen, not only mature antibody producing B cells are produced, but immature B and T memory cells are also produced, which can respond on a second exposure to the antigen (Fig.).

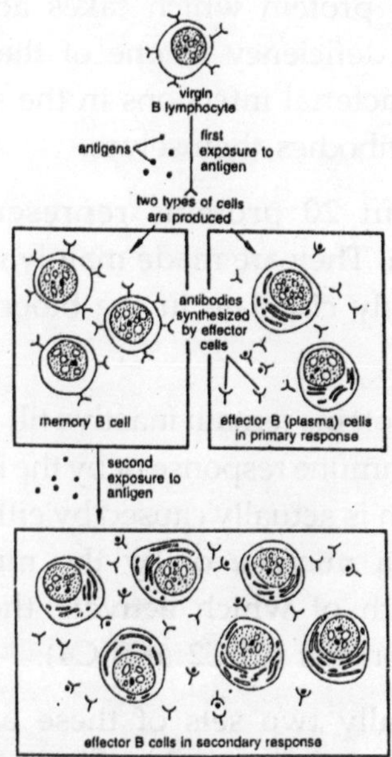

Cellular basis for greater speed and intensity of a secondary antibody response.

About 10,000 different antibodies are known to exist in a human body at a given time (there are 10^{12} lymphocytes with 10^6-10^8 specificities). Obviously thousands of plasma cells throughout the circulatory system are in operation for the production of antibodies at any given time. ' About 90-95% of the protein produced in plasma cells gives rise to antibodies, which are sometimes stored in large quantities in these cells, and can be visible as crystals within the cell. With the introduction of an antigen inside the body, the antibody-producing plasma cells increase in number, their RNA content increases and the entire protein synthesizing machinery is activated for use in antibody production.

Complement System for Antibodies : The action of antibodies is often complemented and amplified by some interacting soluble proteins, which constitute a complement system. One of the central or pivotal component of this complement, is C3 protein which takes active part in this phenomenon. Any deficiency in one of the C3 components leads to repeated bacterial infections in the same way as due to deficiency of antibodies themselves.

There are about 20 proteins representing the above complement system. They are made mainly in the liver. These proteins continuously circulate in the blood stream and the lymphatic system.

Most of these proteins remain inactive till they are triggered into activity by an immune response or by the invading microbe. The above activation is actually caused by either the antibodies on the surface of a microbe or by the microbial envelope polysaccharide, both of which activate the so-called early complement components (Cl,C2 and C4).

There are actually two sets of these early complement components, ' one belonging to the classical pathway, that is activated by IgG or IgM antibodies and the other belonging to the alternative pathway activated by microbial envelope polysaccharide (Fig.). The components of both these pathways activate C3, and the cleavage of this C3 protein leads to (i) assembly of late complement components (C5-C8) forming membrane attack complexes at the microbial cell membrane, that are found as pores or holes (Fig.) and (ii) recruitment of various WBCs. The membrane attack complexes can cause lysis of RBCs.

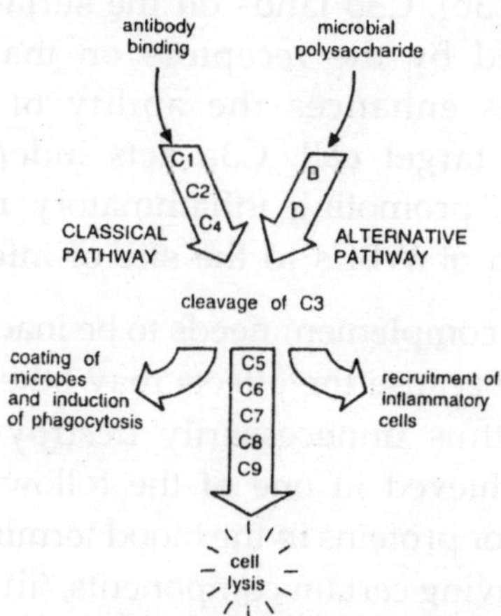

Steps in complement activation by the classical and alternative pathways (C1-C9 and factors B and D are the reacting components of the complement system; Cl, C2, C3, Ç4, B and D are early components, C5-C9 are late components (redrawn from Alberts et al. 'Molecular Biology of the Cell; 1994)

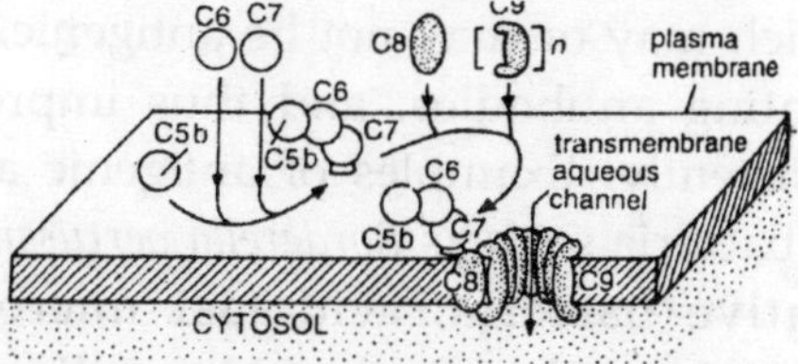

Assembly of late components to form a 'membrane attack complex' (redrawn from Alberts et al. 'Molecular Biology of the Cell', 1994).

The early components and C3 are known to be proenzymes, each of which is sequentially activated by proteolytic cleavage of the preceding proenzyme in the sequence. This cleavage at each step generates a serine protease, which then cleaves the next proenzyme in the sequence, thus establishing an amplifying proteolytic cascade. Consequently each molecule, that is activated in the beginning leads to the production of many active compounds including 'membrane attack complexes' on the microbial membrane.

On cleavage, C3 may give a small fragment (C3a), and a large fragment (C3b). C3b lands on the surface of target cell and is recognized by the receptors on macrophages and neutrophils. This enhances the ability of these cells to phagocytose the target cell. C3a acts independently as a diffusible signals, promoting inflammatory response in the form of migration of WBCs to the site of infection.

The activated complement needs to be inactivated, after its function is over. Because the effects may otherwise spread to adjoining cells, thus unnecessarily destroying them. This inactivation is achieved in one of the following two ways : (i) Specific inhibitor proteins in the blood terminate the cascade by binding or cleaving certain components, (ii) Many activated components in the cascade are unstable so that unless they bind to an appropriate component in the chain or a nearby membrane, they become inactive.

Function of Adjuvants and Immunosuppressants : Several types of treatments modify the immune response in animals. Adjuvants, which may or may not be antigenic, increase the level of circulating antibodies, and thus improve immune responses consistently. Examples of antigenic adjuvants are gram negative bacteria such as *Borderella pertussis*, endotoxins of gram negative bacteria, acid fast micro-organisms, phytohemagglutinins and antibodies. Non-antigenic adjuvants include potassium aluminium tartarate, calcium phosphate, mineral oil, lanolin, surface active agents, calcium alginate, collodion or acrylic particles and polynucleotides. Often a mixture of several such substances is used as a complete adjuvant. Freunds complete adjuvant is such a mixture which contains killed mycobacteria, an oil and an emulsifier.

In contrast to adjuvants, there are immuno-suppressants, which include several physical agents (such as X-rays or gamma rays), chemical reagents and biological agents, that suppress the immune response in animals. Some corticosteroids, purine analogs, folic acid antagonists and alkylating agents are potent immunosuppressants. The enzymes ribonuclease and asparaginase also suppress immune responses.

The suppression of antibody production can be brought about by any one of the following five general biological means: (i) There may be a deficiency in the ability of cells to synthesize gamma globulins (components of immunoglobulins). This deficiency is called hypogammoglobulinemia or agammoglobulinemia. (ii) There may be a competition between different antigens. Thus, one antigen may suppress the immune response due to another antigen, (iii) Administration of an antilymphocyte serum (ALS) causes loss of lymphoid cells and reduction in the antibody response to antigens. Antilymphocyte serum is prepared by immunization of a heterologous species with cells from lymph node, thymus or spleen. (iv) There is a feedback inhibition of immunoglobulin production. Passively-administered antibodies often inhibits the production of new antibodies, (v) Some systems might develop immunologic tolerance. The system may have tolerance to a specific antigen and would not produce its antibodies.

Immunoglobulin Structure

There are five major types of antibodies or immunoglobulins, namely IgA, IgD, IgE, IgG and IgM which may differ in molecular weight in the range of 150,000 to 720,000. Several of them can be further subdivided into subclasses like IgGl, IgG2, etc. Immunoglobulin G (IgG) is the

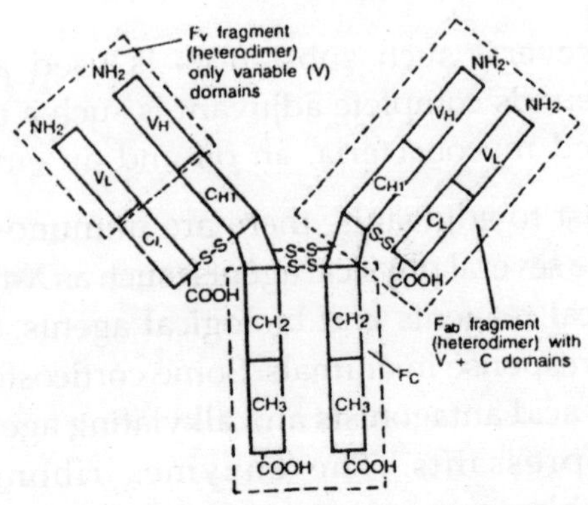

A model of the structure of a typical human antibody molecule, composed of two light and two heavy polypeptide chains (note carefully the presence of disulphide S-S bonds and carbohydrate moieties; also note constant regions, Cl and Ch and the variable regions, Vl and Vh).

commonest and simplest of all immunoglobulins. It is a tetramer containing two identical high molecular weight or heavy chains and two identifical low molecular weight or light chains. Each heavy chain is made up of about 500 amino acids, the light chain having only about half this number. These four chains are linked covalently by disulfide bonds in a Y-shaped structure (Fig.). Each Ig chain consists of two distinct parts : (a) C-terminal (carboxyl- terminal) constant part or C-region and (b) N-terminal (amino terminal) variable part or V-region. For each C-region, there can be several V sequences $(C_{\alpha}, C_{\mu}, C\gamma, etc.)$.

Separate sets of V sequences are known for heavy chains (V_H) and light chains (V_L); (V_k) = for kappa and V_{λ} = for lambda). The different V sequences have fair degree of homology only in some of the variable portions. There are distinct genes for the synthesis of C and V regions of polypeptides.

The Y-shaped structure of antibody can be divided into following functional domains (Fig.) : (i) antigen-binding domains—Fv and Fab fragments and (ii) effector domain—Fc fragment, which helps in elimination of antigen. Fv fragment is a heterodimer, consisting of the variable domains of the heavy chain and the light chain (V_H and V_L). Fab fragment is also a heterodimer and consists of Fv fragment + C_{H1} and CL (constant domains of heavy and light chains).

In other words, Fab consists of the entire light chain and the V_H and C_{H1} domains of heavy chain. Fc fragment consists of a homodimer with C_{H2} and C_{H3} domains of both heavy chains. The entire antibody can be enzymatically cleaved at the hinge region into Fab and Fc fragments. The identification of these fragments of antibody has allowed the application of protein engineering and genetic manipulation for the production of novel antibodies without the use of hybridomas or animal cells.

There are five types of heavy chains, designated as α (alpha), δ (delta), ε (epsilon), γ (gamma) and μ (mu) in IgA, IgD, IgE, IgG and IgM respectively. However, there are only two types of light chains, designated as V_κ (kappa) and V_λ (lambda). All five types of heavy chains occur in alternate forms containing either V_κ or V_λ chains, thus resulting in ten different combinations.

Immunoglobulin A (IgA) is the major antibody found in secretions (tears, mucus and saliva), which is the first defense against invading bacteria and viruses. The functions of IgD and IgE are not known, although it is known that IgE differs from other immunoglobulins in being thermolabile, other immunoglobulins being stable at 50-60°C for half an hour to 4 hours. IgM, which appears early in response to many invading pathogens is bigger than IgG (which replaces IgM later) and constitutes about 5-10% of the total immunoglobulins in the adult human, making about 125 ± 45 mg per 100 ml of serum. Some of the properties of human immunoglobulins are listed in Table.

Some Properties of Different Types of Human Immunoglobulins.

Characteristics	*IgG*	*IgA*	*IgM*	*IgD*	*IgE*
Svedberg units	6-7S	6-11S	19S	6.1S	8.2S
Molecular weight	150kd	160kd	900kd	185kd	200kd
Carbohydrate (%)	3	8	12	13	12
Per cent total Ig concentration	75-85	10-15	5-10	1	0.002
Concentration (mg/100 ml of serum)	1275	225	125	3	1

Three Regions of Hypervariability

Complementarity Determining Regions or CDRs, in Each Heavy and Light Chain, Forming Antigen Binding Sites : The variability of the variable regions of light and heavy chains of antibodies is largely attributed to the three small hypervariable regions in each of the four chains. The remaining part of the variable region is known as framework regions, that are not really variable, but are constant. Only 5-10 amino acids in each hypervariable region form the antigen binding site (Fig.). In view of this, the binding antigen or antigenic determinant (epitope) is also known to be small (fewer than 25 amino acid residues). These three hypervariable regions are also called complementarity determining regions (CDRs).

Antibodies Glycosylation Process : Antibodies are also glycosylated, mostly carrying carbohydrate moieties in C_{H2} region. The carbohydrate component of the immunoglobulin is an oligosaccharide with variable mono-saccharide units. It may include simple mannose-rich units as in IgM or complex N-acetyl-lactosamine rich units as in IgG (Fig.)- In IgG, the carbohydrate portion is about 2.5% of the total molecule and it is equally divided between the two gamma heavy chains, which are linked through amino acids, aspartic acid and threonine residues in the polypeptide chains. Mostly there is no glycosylation in Fab fragments and therefore glycosylation generally has no effect on antigen-binding property of antibody. It, however, does influence effector function controlled by F_C fragment.

Antibody Productions Genetics and Molecular Biology : Antibodies or immunoglobulins are coded by genes, which undergo DNA rearrangements during cell differentiation, thus suggesting that although the genome of an orgainism remains constant in different cells and in the germline, it does undergo a change in B and T cells for the immunoglobulin genes in mammals. As discussed earlier in this chapter, it has also been shown that each heavy or light chain consists of one or more constant regions (coded by C genes) and a variable region (coded by V genes), so that each polypeptide is derived from two genes. This has led to the formulation of two genes one polypeptide hypothesis in contrast to the popular one gene, one polypeptide hypothesis applicable for most gene systems.

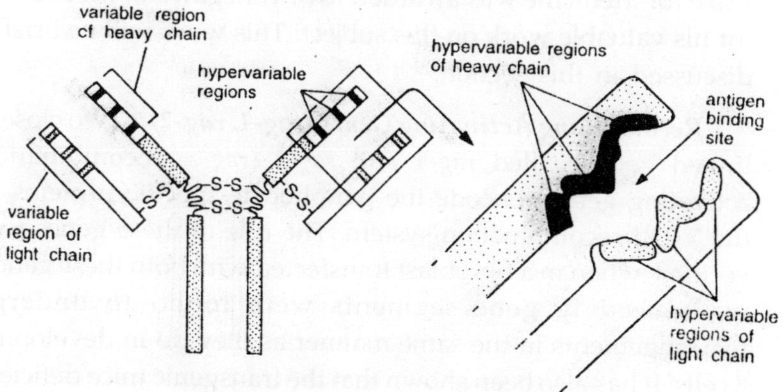

Three antibody hypervariable regions (also called complementarity determining regions or CDR), each in heavy and light chain, forming antigen-binding site of an antibody molecule (redrawn from Alberts et al. "Molecular Biology of the cell", 1994).

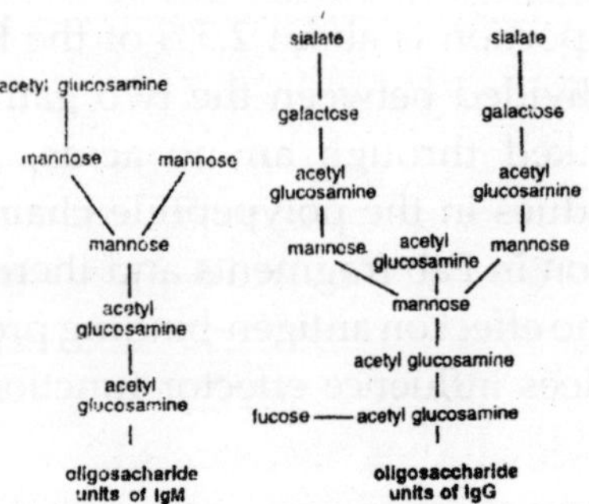

Details of the structure of carbohydrate moieties (oligosaccharide units) found in antibodies IgM and IgG.

The immunoglobulin gene system is unique in the sense that thousands of different antibodies are synthesized in different cells, which are supposed to have the same DNA content. Surprisingly, we do not have as many genes as the variants of antibodies in an organism, but instead sequences coding for different types of antibodies are actually derived in different cells through rearrangement of the master DNA sequence available in the germline. This has been achieved by a remarkable mechanism discovered during 1970's and 1980's. In view of the significance of this discovery, the 1987 Nobel Prize for Medicine was awarded to S. Tonegawa of MIT (USA) for his valuable work on this subject. This work will be briefly discussed in this section.

Recombining Activating Genes (rag-1, rag-2) **:** Two closely linked genes called *rag-1* and *rag-2* (*rag* = recombination activating genes) encode the lymphocyte-specific proteins of the V(D)J recombination system. The role of these genes was verified, when in a fibroblast transfected with both these genes, introduced Ig gene segments were found to undergo rearrangements in the same manner as they do in developing 8 cells. It has also been shown that the transgenic mice deficient in either of these two genes (*rag-1, rag-2*) were unable to initiate V(D)J rearrangements. Consequently they did not have functional B or T cells.

Ig genes in Germline and Distinguished Cells : By using cDNA probes for different regions of an immunoglobulin, it was possible to demonstrate that in the antibody producing cells the DNA sequences coding for C and V regions of an Ig gene are very close to each other, though not contiguous. On the other hand, in the germline these sequences are separated from each other, sometimes by long distances. This was expected because in earlier studies the same C region was found combined with different V regions in the same individual. Later experiments showed that C-C splicing at the DNA level is responsible for generating the coding genes for both light and heavy chains. It was also shown that although the genes for an immunoglobulin molecule may be found on more than one chromosome, but cutting and rejoining of DNA sequences takes place only within a chromosome and not across chromosomes. In addition to C and V sequences, both light and heavy chain coding sequences have another sequence called J (junction or joining sequences). In heavy chain sequence, in addition to C, J and V sequences, another short sequence D (diversity) is also found.

V-regions by DNA Splicing

Splicing and rearrangement of DNA sequences take place in antibody producing cells to create a large variety of Ig genes. As an example, we will discuss DNA splicing only for the heavy chain.

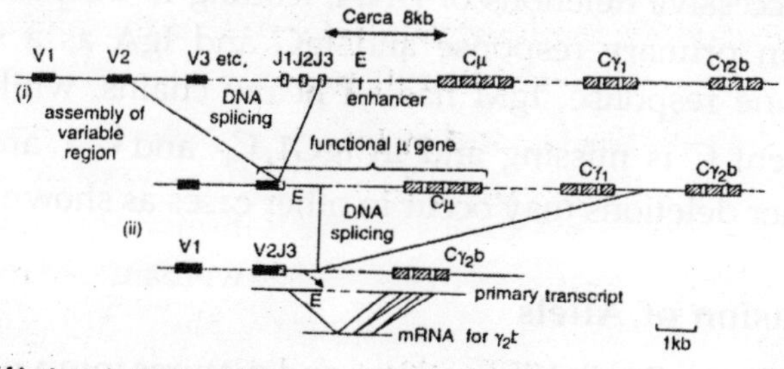

(A) steps in DNA - splicing

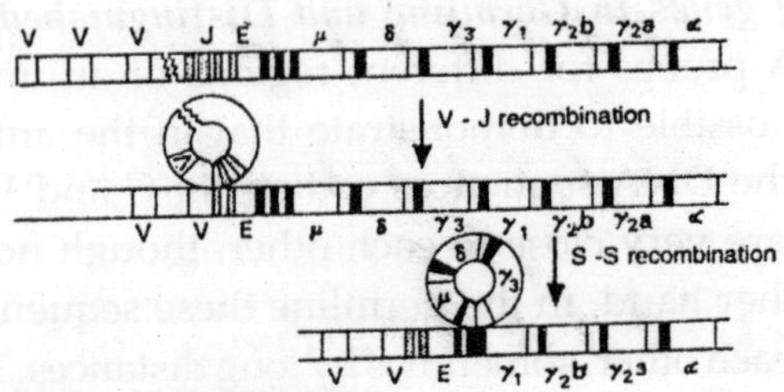

(B) model of heavy - chain switching

V-J joining process involved in making a K light chain in mouse.

There are two rearrangements in this case : (a) the first generating the DNA sequence coding for V_H and (b) the second leading to association of a specific V_H with a definite C_H sequence. Using molecular probes in different kinds of cells, steps involved in rearrangements leading to association of different V regions could be worked out as shown in Figure. In IgM sequence, it is shown that in embryo or germline cells, V_H and C_H regions have a small region of J sequence. Between V and J sequences, presence of D sequence was also shown. Although there is diversity in V, J and D sequences (C region is constant), but variation is also provided by the manner in which DNA splicing and rearrangement is brought about.

V-C Connection Via DNA Splicing : As indicated above, the second phase of DNA rearrangement involves association of a V segment with a C segment. It has been shown that the same V segment can attach with different C regions, thus providing for H-chain class switching (Fig.). This is achieved by successive deletions of DNA, leading to the production of IgM in primary response and IgG and IgA as a secondary immune response. IgM has all heavy chains, while in IgG_3, segment C is missing and in IgG1, C_μ and $C_{4\gamma^3}$ are missing. Further deletions may occur in other cases as shown in Figure.

Exclusion of Allels

When a B cell differentiates and matures into a plasma cell, it synthesizes solitary types of light and heavy chains. This happens despite the presence of two loci (one paternal and one maternal) for heavy chain and four loci (two paternal and two maternal) for light chain in a diploid cell.

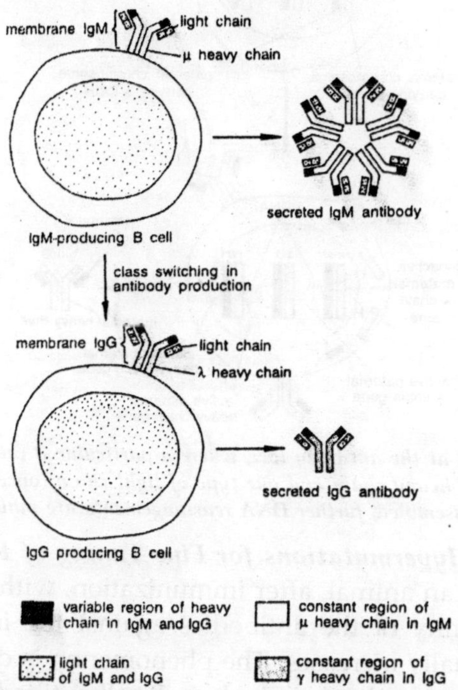

Attachment of the same Vh region to different Ch regions for Ig class switching, during antibody development

This phenomenon, where some of the loci are excluded from being expressed is known as alletic exclusion and is unique to antibody gene expression (Fig.). In the expression of antibody genes, V_H genes are rearranged first, and then the genes for k light chain segments are assembled.

The λ light chain genes are assembled only if rearrangements at the V_κ locus fail to produce a functional K light chain. The rearrangements of V_H and V_κ genes thus inhibit any further rearrangement.

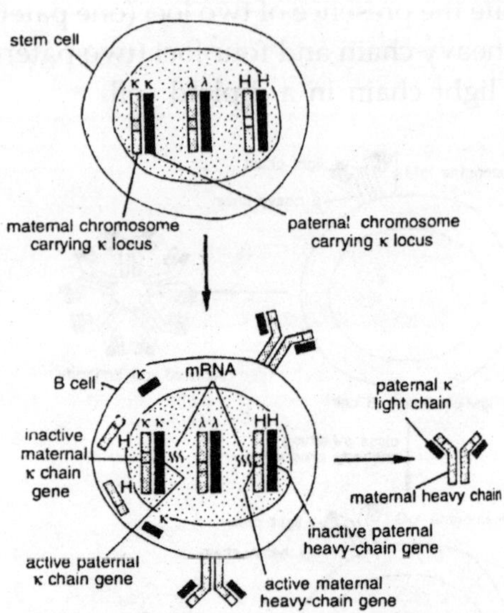

Allelic exclusion at the antibody loci, where a particular B cell synthesizes only one type of heavy chain and one type of light chain (once a functional gene is assembled, further DNA rearrangements are inhibited).

Somatic Hypermutations for Fine Tuning of Response of Antibody : In an animal, after immunization, with passage of time, the affinity of the antibodies against the immunizing antigen gradually increases. The phenomenon is described as affinity maturation. It occurs only in B cell antibodies and not in T cell receptors. This is due to accumulation of point mutations in DNA sequences coding for the variable regions of heavy and light chains. These mutations occur long after the DNA rearrangements occur, and also after the B cells are stimulated by either an antigen or by helper T cells to generate memory cells in the activated centre (also called germinal centre) of a lymphoid follicle.

These mutations occur the rate of about one mutation per V-region coding sequence per cell generation. This rate of mutation is very high and is one million times greater than the spontaneous rate. The process, is therefore, called somatic hypermutation. As a result of repeated cycles of somatic hypermutations followed by antigen-driven selection, antibodies of increasingly high affinity are produced providing protection against harmful antigens.

Functions of Interferons

Interferons, which were discovered for the first time in 1957 by two British scientists (Alick Isaacs and Jean Lindermann), are proteinaceous substances (MW 18,000 to 100,000) produced inside the body for defense against virus infection. They are also produced in response to many non-antigenic chemicals including polysaccharides, endotoxins, double stranded DNA, RNA, etc. Unlike immunoglobulins, interferons are stable for several hours at room temperature at pH 2.0.

Two mechanisms for the control of virus infection by interferons have been suggested :

(i) interferons affect the cellular membranes, so that the virus particles are not disseminated to the other cells;

(ii) interferons control the multiplication of virus particles by inhibiting their protein synthesis. It is believed that the interferons set off a number of enzymes in the cell which can stop the virus using the cells protein synthesizing machinery for its own protein synthesis. It has also been observed that ribosomes from interferon infected mammalian cells do not form polysomes with viral RNA, but when they are exposed to host (mammalian) RNA, they readily form polysomes and protein synthesis is initiated. It shows that ribosomes in the presence of interferon can recognise specific mRNA and regulate the overall protein synthesis.

Because of their inhibitory nature, the interferons are being tested for their effectiveness in controlling cancer, though with only limited success. In some cases, interferons inhibit tumour cell growth and stimulate immune cells to destroy cancerous cells. With some exceptions, the tumours that respond to interferon treatment are also responsive to chemotherapeutic agents. Many companies in U.S.A., Japan, U.K., W. Germany, France and Switzerland are working towards large-scale production of interferor.

Developing Vaccines and Immunization

Backgroun and Growth of Vaccine Development : Vaccine production and immunization or vaccination is the most important application of immunotechnology, as far as human welfare is concerned. The first experiment in immunization was performed by Louis Pasteur, a French chemist turned biologist, on July 6, 1885, when he treated a young boy ' against rabies. He extracted fluid from the spinal cord of a rabid dog and injected it in small amount to a boy, who was later bitten several times by a rabid dog. Apparently the extract from the spinal cord had stimulated the production of antibodies against the rabies virus. The science and application of the vaccine technology, however, made significant progress in the second half of the 20th century, so that the period 1950-70 is considered to be the golden age for vaccine development. During this period, vaccines for poliomyelitis, measles, mumps and rebella were developed. In majority of the vaccine recipients there is no significant ill effects and mortality rate among vaccines varies from about one per 200,0000 to one in 3,000,000. Thus the vaccines are quite safe and cost effective. The protection given by measles vaccination is documented in Table.

Benefits of Vaccination against Measles (data taken from A.K. Galazka and others, Bull WHO 1984, 62 : 357).

Adverse reaction	*Incidence (no. per 100,000 cases) following*		*Ratio Infection vacci-nation*
	Infection	*Vaccination*	
Pneumonia	5000	1	5000
Convulsions	720	60	12

Contd...

Adverse reaction	*Incidence (no. per 100,000 cases) following*		*Ratio Infection vaccination*
	Infection	*Vaccination*	
Brain damage	280	0.1	2800
Early death	10	0.02	500
Chronic infection	2	0.075	26

However, production of vaccines against certain important infectious agents has been difficult. The problems involved in the production of vaccine in such cases include the following : (i) Antigenic variation or presence of many serotypes as in influenza, rhinovirus and in human immunodeficiency virus (HIV), (ii) Integration of viral DNA/cDNA in the host cell genome as in hepatitis B and retrovirus. (iii) Large animal reservoirs as in influenza and human HIV. (iv) Infection transmitted by cells which may or may not express viral antigen as in HIV. (v) Crucial cells of the immune system are infected as in HIV. If a pathogen has most or all of these characteristics, it is difficult to produce an effective vaccine. The HIV causing AIDS at present seems to fit this description.

Vaccines' Nature

In broader sense, the term vaccine is used for any biological material injected in the body to stimulate the active immunity. Immunization of the body through vaccination is generally done for the prevention of disease. The vaccine is administered in advance so as to give the body time to set active immunity, before invasion by the pathogen occurs. An objective of vaccination is to stimulate the production of antibodies without the person actually having to develop the disease. The vaccines may be of following types :

Organisms of Living Kind (with Attenuated Virulence) as Vaccines : Some vaccines involve use of organisms, whose virulence is greatly attenuated or diminished. These organisms are unable to produce disease but induce immunity when injected in the host. There are at least five different ways of producing attenuated strains of pathogens :

(i) Many successive passages of pathogen in some animals other than the usual host of the pathogen or in cultured cells,

(ii) Selection of mutants of low virulence,

(iii) Treatment of pathogens with chemicals,

(iv) Cultivation of pathogens under unfavourable conditions such as high temperature,

(v) Deletion or insertion of genes leading to attenuation of pathogens as tried for vaccinia virus.

Some examples of vaccines prepared by the attenuation of virulence are given in Table

Vaccines Prepared by Attenuation of Virulence

Disease/pathogen	*Method of attenuation*
1. Rabies	Cultivation in chick or duck embryo
2. Yellow fever	Cultivation in chick embryo
3. Smallpox	Animal passage (calf); genetic engineering
4. Polio	Mutant strain (sabin oral polio vaccine)
5. B.C.G. (*Bacillus* of Calmette Gueries)	Cultivation of organism on media containing bile

Contd...

Disease/pathogen	*Method of attenuation*
6. *Bacillus anthracis* (Anthrax)	Cultivation at 40-50'C

The concept of recombinant 'vaccinia' virus as a live vaccine gained momentum in 1990's due to its earlier effective use as a smallpox vaccine. Its host range makes it suitable for livestock also. Following attributes of *Vaccinia* (used as an expression vector) made it attractive for vaccine development : (i) stimulation of humoral and cell-mediated immunity, (ii) low cost, (iii) heat stability, (iv) simple method of administration, (v) visible proof of successful vaccination and (vi) long duration of effectiveness after a single dose.

Genetically Engineered Viruses as Vaccines : Although in the past attenuated strains of vaccinia virus were obtained by serial virus passage, genetic engineering methods were used for this purpose in 1980's and 1990's. Deletion of several genes (e.g. genes for thymidine kinase, a growth factor, hemagglutinin, 13.8 kD secreted protein, and ribonucleotide reductase) decreases its virulence. Similarly, insertion of some lymphokine genes (e.g. genes for interleukin-2 or IL-2) also lead to decreased virulence. While reducing virulence or pathogenicity, immune response due to recombinant protein (coded by gene inserted in the virus for developing specific vaccine) must be increased.

More than one genes can also be stably integrated into attenuated vaccinia virus for developing recombinant virus to be used as a vaccine against multiple pathogens. A glycoprotein gene of rabies virus has been integrated and the resulting rabies virus vaccine has been field tested in Europe and U.S.A. giving promising results. Other vaccines against HIV are being developed using the gene for HIV-1 envelope glycoprotein. In future, the widespread use of *Vaccinia* as a live vaccine will depend on improving safety, while achieving higher response to the recombinant protein. Dead organisms as vaccines. In some cases, the dead bodies of the pathogen, when infected into a host, stimulate the production of antibodies without causing any infection. This method has been used in immunizing against typhoid fever, whooping cough, blackleg in cattle and a few other diseases.

Use of Bacterial Toxins and Toxoids as Vaccines **:** Erotoxins of most pathogenic bacteria are highly antigenic; they stimulate the production of antibodies (antitoxins) in animal or human body. These exotoxins are highly unstable. They lose toxicity during storage and are converted gradually into toxoids. Toxoids retain the antigenic power but they are not toxic. Therefore, they can be used in producing immunity against corresponding toxins. The transformation of toxins into toxoids is accelerated by heat, formaldehyde and other chemicals. Alum-precipitated toxoids are highly antigenic. Toxoids have been used to develop immunity against several diseases including diphtheria, tetanus; gangrene, etc.

Immunizing Sera as Vaccines **:** An immunizing serum (plural 'sera') is the actual blood serum of a person or animal and contains an immunizing material or antibody. It is generally obtained by injecting a toxoid into an animal (usually a horse), permitting establishment of active immunity in the animal, and then withdrawing some of its blood. The corpuscles are removed from serum, which is then standardized as to the strength of antibodies and stored in sterile tubes or bottles.

Immunizing sera are used mostly for cure rather than prevention. They may be given for prevention after exposure has taken place, but their immunizing effects are too shortlived to give any benefit if administered much in advance of the disease. As a curative agent, the action of immunizing serum is very rapid. Immunizing serum has been used in the cure of diphtheria and tetanus.

Vaccines of Defined Chemical Nature : Since the chemical nature of many antigens is now known, it has been possible to produce these antigens either by biological or by chemical means. Perhaps the first defined molecular species to be used as a vaccine were the polysaccharide vaccines for *Neisseria meningitidis* (meningitis) and *Streptococcus pneumoniae* (pneumonia). The pure polysaccharide preparations from the bacterial capsule could be used as an antigen in adults or children of over two years of age. Addition of some other chemicals or biological principles often increases the potency of the polysaccharide antigen. For example, *Haemophilus influenzae* (influenza) B polysaccharide is very poor in inducing antibodies in the susceptible host on its own.

However, its antigenic property is greatly enhanced if it is mixed up with diphtheria toxoid. Proteins and nucleic acids have also been identified as immunogenic molecules. An outer membrane protein-polysaccharide complex from meningococci bacteria in association with aluminium phosphate or hydroxide elicits immunogenic responses, so that the mixture can be used as a vaccine. Antigens produced using cloned genes, synthetic peptides and proteins along with certain adjuvants are also often used as vaccines. Nucleoprotein of influenza virus can also be used as a universal vaccine for influenza.

The advent of recombinant DNA and monoclonal technologies coupled with the recent advances in immunology, has opened a new era into the production of vaccines. It seems that isolated immunogens can now be produced in unlimited quantities and vaccine development can proceed without hindrance. Several vaccines based on the new technologies have been licensed for commercial production. Some of these include the following : (i) Yeast-produced vaccine for hepatitis B virus based on cloning the DNA sequence for the antigen; (ii) *Escherichia coli* vaccines for pigs with extra plasmids encoding the antigen for K88 strain of *E. coli;* (iii) Monoclonal antibody against K99 strain of *E. coli* for protection of calves.

Vaccines of Defined Chemical Nature : Since the chemical nature of many antigens is now known, it has been possible to produce these antigens either by biological or by chemical means. Perhaps the first defined molecular species to be used as a vaccine were the polysaccharide vaccines for *Neisseria meningitidis* (meningitis) and *Streptococcus pneumoniae* (pneumonia). The pure polysaccharide preparations from the bacterial capsule could be used as an antigen in adults or children of over two years of age. Addition of some other chemicals or biological principles often increases the potency of the polysaccharide antigen. For example, *Haemophilus influenzae* (much used) polysaccharide is very poor in inducing antibodies in the susceptible host on its own.

However, its antigenic property is greatly enhanced if it is mixed up with diphtheria toxoid. Proteins and nucleic acids have also been identified as immunogenic molecules. An outer membrane protein-polysaccharide complex from meningococci bacteria in association with aluminium phosphate or hydroxide elicits immunogenic response, so that the mixture can be used as a vaccine. Antigens produced using cloned genes, synthetic peptides and proteins along with certain adjuvants are also often used as vaccines. Nucleoprotein of influenza virus can also be used as a universal vaccine for influenza.

The advent of recombinant DNA and monoclonal technologies coupled with the recent advances in immunology, has opened a new era into the production of vaccines. It seems that isolated immunogens can now be produced in unlimited quantities and vaccine development can proceed without hindrance. Several vaccines based on the new technologies have been licensed for commercial production. Some of these include the following: (i) Yeast produced vaccine for hepatitis B virus based on cloning the DNA sequence for the antigen. (ii) *Escherichia coli* vaccines for pigs with extra plasmids encoding the antigen for K88 strain of *E. coli*. (iii) Monoclonal antibody against K99 strain of *E. coli* for protection of calves.

9

Biotechnology in Medicine

Generally medicines/drugs are prescribed for treatment of specific diseases based on the concept "one-medicine-fits all". In other words medicines are given for a disease without looking into the constitution of the patient, except in some cases, where a physician may enquire or examine about allergenicity of the patient to sulphur drugs or some antibiotics. However, it is now known that the response to a particular drug may differ in different patients suffering with the same disease. The study of how genetic differences of patients may influence the variability in the response to specific drugs is described as pharmacogenetics.

Efficacy of major drugs used for several important diseases has been tested and the desired response of patients to these drugs varies from 25% to 80% (Table). In the year 1994, in USA alone, 1.8 million people were hospitalized and another 100,000 deaths occurred for adverse drug effects. Therefore, drug response should be known before a particular medicine is prescribed and patients be classified as responders and non-responders. The responders may also be classified in two groups, those who derive benefit from drug and those exhibiting side effects. This will allow an estimate of benefit-risk ratio, permitting a judicious decision for prescribing a specific drug.

It has been recognized that the presence or absence of response to a drug, and also its adverse side effects may sometimes be attributed to reasons like mis-dosing, drug-drug interactions, allergy or medication error. However, it has been shown that efficacy and toxicity (side effects) of many medications depend on genetic polymorphisms in patients for drug-metabolizing enzymes, transporters, receptors and other drug targets.

Degree of Response of Patients to a Major Drug for a Selected Groups of Therapeutic Areas.

Therapeutic area	*Efficacy rate (%)*
Alzheimer's	30
Analgesics (Cox-2)	80
Asthma	60
Cardiac arrythmias	60
Depression (SSRI)	62
Diabetes	57
HCV	47
Incontinence	40
Migraine (acute)	52
Migraine (prophylaxis)	50
Oncology	25
Osteoporosis	48
Rheumatoid arthritis	50
Schizophrenia	60

These genes may also encode proteins involved in drug absorption, distribution, metabolism and elimination. Many of the genes determining response to drugs are non-disease and non-gene specific and are being studied in some detail. These studies have also shown that the drug response is polygenic in many cases, thus making their clinical study difficult. A study of these genetic polymorphisms is providing a strong scientific basis for optimising drug therapy on the basis of each patient's genetic constitution. These studies have also been facilitated through genomics research making available whole genome sequences and their annotations into specific functions. Human genome project, coupled with functional genomics and high throughout screening methods is providing powerful new tools for elucidating polygenic components of human health and disease. This area of pharmacogenetics involving genomics research which will allow drug makers to tailor a therapy to the individual's specific needs, is often described as pharmacogenomics.

It has been described as the marriage between functional genomics and molecular pharmacology. The significance of this area is evident from the fact that a new journal "Pharmacogenomics" has been started by the Nature Group of Journals. Thus pharmacogenomics refers to entire spectrum of genes that determine response and sensitivity to individual drugs, whereas pharmacogenetics refers to narrow spectrum of inherited differences in drug metabolism and disposition. Such a distinction between pharmacogenomics and pharmacogenetics is however, arbitrary and the two terms are often used interchangably.

The significance and economic value of a simple predicted drug response, which will provide information on the likelihood of efficacy and safety of a drug for an individual patient will change the practice and economics of medicine. This will make it necessary for a physician to prepare patients' profiles, and based on these profiles prescribe different discrete medicines concurrently for different patients with same disease. Further, the same disease phenotype may result due to different interactions of inherited metabolic variations. The science of pharmacogenetics is providing tools to classify the heterogeneity of disease. Thus, pharmacogenetics and/or pharmacogenomics have already become an important area of biotechnology research to examine heterogeneity of disease and individual's responses to medicine. In this chapter, we propose to discuss examples that illustrate the current status of pharmacogenetic and pharmacogenomic research and also discuss the future prospects of this emerging and fascinating area of biotechnology research.

Pharmacogenetics and Personalized Medicine : As discussed above, the post-genomic era has started with a promise of tailoring the practice of medicine to the individual. This is sometimes described as personalized medicine. However personalized medicine and pharmacogenetics do not mean the same thing, because pharmacogenetics is only a part of personalized medicine, which includes the use of risk algorithms, molecular diagnosis, targeted therapies and, of course, pharmacogenetics (Fig,). The advantages of personalized medicine are listed in Table.

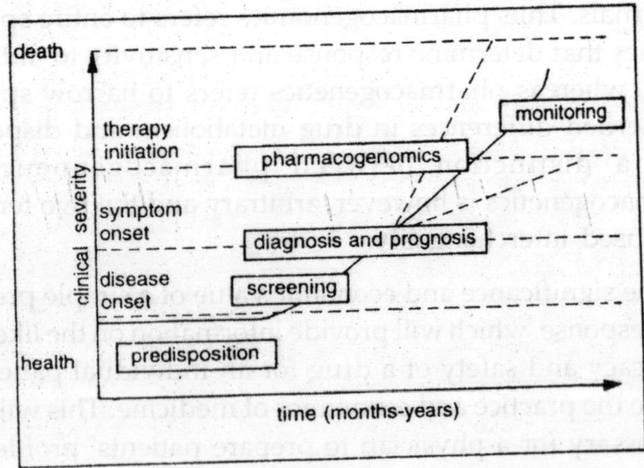

Different stages of a hypothetical disease, where personalized medicine can be useful at each stage; the type of markers and their commercial applications will vary for specific diseases.

Advantages of Personalized Medicine to Patients and Clinicians

1. Higher probability of desired outcome with a drug
2. Low probability of untoward side effects
3. Preventive strategies
4. Focussed therapies
5. Reduced costs
6. Better health and better healthcare

Genetics and Genomics in Medical Practice (Clinical Relevance) : In medical practice, a physician diagnoses the disease and prescribes a treatment on the basis of symptoms/ physical signs and appropriate tests (blood, urine, X-rays, NMR, etc.). However, all patients suffering with a disease may not have same symptoms. Similarly same treatment may not suit all patients suffering with a particular disease. Therefore, pharmacogenetic and/or pharmacogenomic intervention are needed in both cases.

Diagnosis, Mechanism of Disease and Response of Patients to Medicines : Generally, a patient may not exhibit all symptoms of a particular disease. The underlying mechanism for the same disease may also differ in different patients and could be unknown in many cases. It has also been shown in some cases, that mutation in different genes may lead to clinically identical patients. For instance, mutations in genes *APP*, *presenilin 1* and *presenilin 2* lead to clinically indistinguishable forms of Alzheimer's disease. Conversely, mutations at different sites along the *APP* gene may lead to two different diseases, early-onset Alzheimer's disease and recurrent intracerebral haemorrhage. In still other cases, several interacting susceptibility genes, through their variant products, may lead to multiple clinical effects.

It is also possible that two or more than two distinct diseases may have same underlying mechanism, so that same medicine may work for both the diseases. These issues will be illustrated using two examples.:

(i) Alosetron is a medicine prescribed in USA for female patients with diarrhoea-predominant irritable bowel syndrome (IBS), which is a complex disease and a response profile to determine whether a patient will get relief from a medicine will be useful.

(ii) Carbamazepine was earlier used to be a medicine prescribed for treatment of trigeminal neuralgia (a syndrome with intermittent severe lightning bursts of facial pain), but later it was also prescribed to treat various forms of epilepsy. Pharmacogenetic approach will allow all such cases to be dealt more appropriately, through a study of genetic basis of patient's response to medicines.

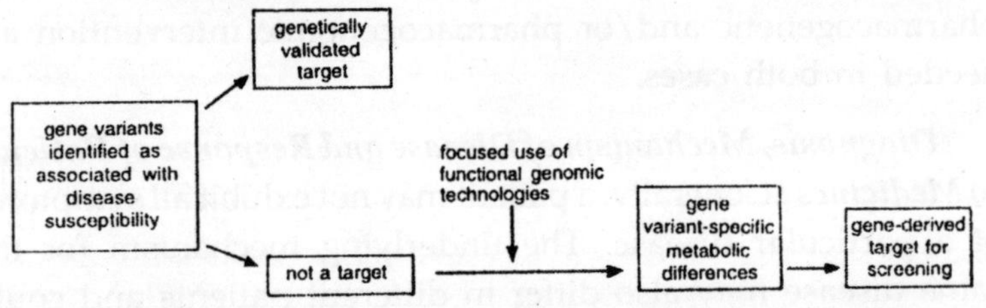

(a) identification of drug target through focused use of genomics

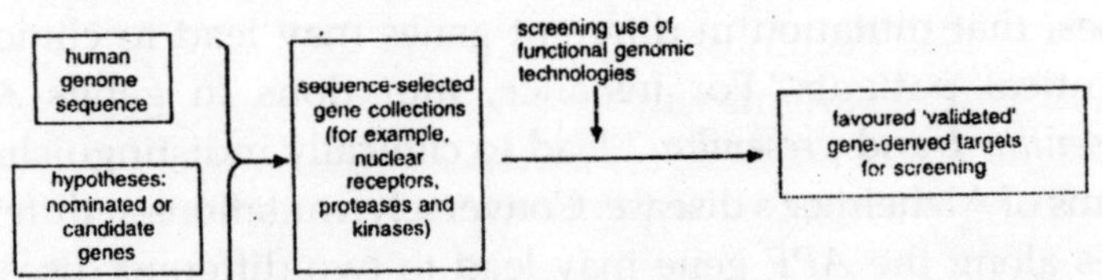

(b) identification of drug target through screening use of genomics

Genetics and genomics as two approaches to identify drug targets, both making use of functional genomics. In the former (genetics) approach, gene is known or identified and product of allele responsible for disease is screened for drugs. In genomics approach, on the basis of homology, genes belonging to similar families are identified, screened and their products are screened as drug targets.

DNA Polymorphism and Treatment of Diseases : It has been shown in many cases that a prescribed medicine may be effective in some patients without any side effects, while in other patients it is ineffective and/or has adverse side effects. Although in such cases, information from clinical trials including data on medicines may be useful, but results of pharmacogenetic research can be more effectively used to prescribe a medicine. It involves collection of information on patient's DNA to determine how a patient is likely to respond to a particular medicine. For instance, polymorphisms in genes encoding cytochrome P450, N-acetyltransferase and other enzymes involved in drug metabolism have a correlation with the concentration of the drug in the blood and therefore can help in prescribing specific doses of the drug to individual patients. There may also be genetic differences leading to differences in responses involving rate of activation of the drug used.

In view of this, enzyme tests have already been replaced in some cases by genetic variants on chips. In rat models of heart disease, it has been shown that the effect of hydralazine (used as an arterial vasodilator) differs in the presence of β and α myosin heavy chain (MHC) alleles. Thus the administration of the same drug had different effects, depending on which allele was expressing. Differential responses of four myosin alleles to a drug bupivacaine (a drug toxic to muscle cells) has also been observed. Therefore, it is expected that in future, genetic variants will be standardized, so that automated read-outs will be possible for each person's predicted response to each medicine.

Although such a DNA-based screening may not help so much in diagnosis, but will certainly help in prescribing specific medication and its specific dose to avoid any side effects. SNP haplotyping will perhaps be the most common method for the study of DNA polymorphism, since SNP alleles will soon become the target for therapeutic treatment (for SNPs, see later).

The earliest significant pharmacogenomics deal was done in July 1997 between Genset and Abbott Laboratories. They agreed to analyze variations in patient's response to particular drug therapies. This deal actually validated the concept of pharmacogenomics; that need to be addressed by any drug-discovery platform. As a part of this agreement, Genset is developing a high-density map of biallelic SNP markers for human genome. This will be followed by identification of markers and genes associated with responses to the administration of specific drugs. Abbott, on the other hand, is developing diagnostic systems for clinical testing of patient's response to these specific drugs.

The diagnostic tests will be conducted at all steps of the development of a drug, i.e. during drug development, clinical trials, regulatory approvals, marketing and sales of the drug. Consequently, both the diagnostic assays and related drugs will be marketed simultaneously and royalties will be payable for both. The example of Genset and Abbott is being followed by other companies also (Table). For instance Genaissance Pharmaceuticals is making its Genthology collection of alleles from clinical populations. This will be followed by studies on expression of these genes/alleles in bacteria, yeast, cell cultures and animal models. Allele targets for drug discovery are thus being identified for breast cancer and cardiovascular diseases, so that customized drugs having higher efficiency and fewer side effects will become available.

Technology Suitable for Clinical Detection of Genetic Molecular Markers

Technology	*Typical markers*
DNA sequencing	SNPs, indels
PCR-based assay	SSRs, SNPs
Hybridization based assay	SNPs, RFLPs, mRNA profile

Contd....

Technology	*Typical markers*
Microarray	SNPs, SBH analysis, mRNA expression profile
Single base primer extension	SNP detection and conformation
Conformation-based assays	SSCPs (single strand conformation polymorphisms)

Use of SNPs in Pharmacogenomics

Single nucleotide polymorphisms (SNPs) are single-base differences in the DNA sequences between individuals and occur throughout the human genome with an average frequency of SNP per 1000 base pairs. As many as 2.8 million SNPs are already available in the database of Celera Genomics, a company that was involved in complete sequencing of the human genome. A map of human genome carrying 1.42 million SNPs was also published in February, 2001, and more SNPs might be added to this map in future.

Such an SNP map will be extensively utilized to realize the promise of pharmacogenomics.

Genome-wide SNP Genotyping

SNP genotyping means to find out the identity of alternative alleles that are known to be present at one or more of a large number of SNP loci in an individual. Such SNP genotyping at an individual locus is generally done through PCR amplification of the segment that contains a particular SNP. When this approach is used for SNP genotyping at millions of SNP loci that are known to be present in human genome, this will involve an exorbitant cost and labour. Therefore, efforts have been made in the past to combine or multiplex PCR reactions to reduce the cost and labour. However, different PCR reactions may not be compatible, thus making this multiplex approach difficult, as the number of loci increases. An alternative strategy to reduce the cost and labour is to design a PCR reaction, which may achieve the goal of multiplexing without using a number of primer pairs in the same reaction. One such approach was to use PCR primers complementary to an interspersed repetitive sequence (e.g. alu sequences), thus permitting amplification of subsets of unique genomic sequences, using a single PCR primer.

Such an approach proved useful under certain circumstances, but will allow amplification of only those unique sequences, which lie adjoining to the interspersed repeat sequence that is used for primer designing, but may not allow genome-wide SNP genotyping. This problem can be largely overcome by using degenerate oligonucleotide primed (DOP)-PCR that was designed for whole genome amplification. This PCR makes use of a partially degenerate oligonucleotide, where approximately one third of the nucleotides in the center of the oligonucleotide are degenerate. In such a DOP-PCR, when 3' end of the primer consists of 6-10 unique nucleotides, effective amplification of a eukaryotic genome (e.g. human and mouse genome) is generally observed.

This approach has recently been successfully utilized for genome-wide SNP genotyping in human, mouse and *Arabidopsis thaliana.* Since whole genome sequences were available for all these three eukaryotic genomes, *in silico* methods were also used for predicting DOP-PCR products and the results of the two approaches were found to be comparable (for details consult PNAS, March 5, 2002, PP-2942-2947). Such a method of genome-wide SNP genotyping should prove useful in reducing the cost of pharmacogenomics approach of personalized medicine.

Two Approaches of Pharmacogenomics : In the field of pharmacogenomics, two approaches are available to utilize SNPs.

(i) In the first approach, often described as 'candidate gene approach', genetic associations are first established between specific genes (susceptibility genes) and diseases. SNPs in these genes are then tested for their associations with the disease in the patients. This approach has already been used for an increasing number of candidate genes (e.g. gene encoding thiopurine methyl-transferase or TPM), a drug metabolizing enzyme; genetic variants for this enzyme are linked to adverse drug reactions.

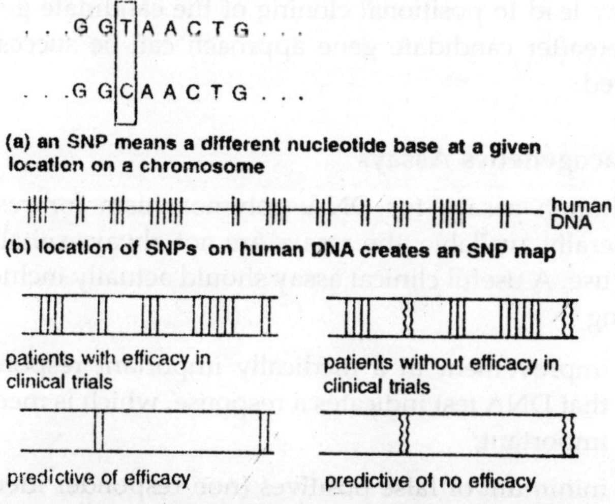

SNPs, SNP maps and their use in pharmacogenetics through association of the efficacy of medicine with SNP profiles (haplotypes).

(ii) In the second approach, called 'linkage disequilibrium (LD) or association mapping approach', SNP maps are used and the genomic region carrying the unknown susceptibility gene is identified through detection of an association of the disease with one or more specific SNP(s). This association is detected through LD.

While in the candidate gene approach, the patient's genotype for SNPs in the candidate gene only will be determined, in the LD approach, every patient will have to be genotyped for all mapped SNPs (~ 1,000,000 SNPs) and then the association of SNPs with the disease and drug response phenotype will be worked out. Since for many diseases, the candidate genes may not be known, LD approach will have to be used, which will be expensive at least initially. With the passage of time, when genomic regions carrying the suspected candidate susceptibility genes become known, only SNPs from specific regions will need to be used for genotyping the patients. This may lead to positional cloning of the candidate gene, so that thereafter candidate gene approach can be successfully employed.

Pharmacogenetics Assays

Although assays for DNA polymorphism for research are generally available, the assays are not always suitable for clinical use. A useful clinical assay should actually include the following :

(i) improvement in a medically important response, so that DNA test indicates a response, which is medically important;

(ii) minimum of false positives (non-responder identified as responder) and false negatives (responder identified as non-responder) so that the efficacy and safety based assays prove useful;

(iii) the results should be interpretable and clinically useful, which can be validated for regulatory acceptance. The technologies that may suit clinical detection of genetic markers are listed in Table.

Pharmacogenomics and Industry

Human genome project and the availability of rough draft of human genome sequence that was published in February,

2001, encouraged the industry to invest in developing tools to make use of pharmacogenomics in medicine. For instance high - resolution maps of biallelic SNP markers and genes associated with responses to drugs are priority areas for companies like Genset, Millenium and Variagenics in USA (Table).

Future of Pharmacogenomics

The future of pharmacogenomics looks bright because differential response of patients with different alleles of a disease gene has already been demonstrated. Thus differential response to drug accounts for (i) variability in clinical trials, (ii) differential side effects and (iii) inconsistencies in disease models. For this purpose, not only nuclear DNA, but even mitochondrial DNA (mtDNA) will be utilized as being done by the company named MitoKor. For instance, it was shown by MitoKor that 60% of patients suspected of suffering with Alzheimer's disease had high level of mutations in mtDNA. Using this information, allele specific drugs will be developed. Variation in drug response in future will certainly be related with variation in SNP profiles.

Selected Companies with Pharmacogenomics Programs (from Nature Biotechnology, Vol. 18 Supplement, 2000)

Company	*Area*
1. Aeiveos Science Group (USA)	Aging-related genes and responses
2. Astra Zeneca (UK)	Population genomic variability
3. CuraGen (USA)	Integrated genomic & pharmacogenomic platform
4. Eurona Medical (Sweden)	Correlation of drug response & genetic profiling
5. Genaissance Pharmaceuticals (USA)	Genetic polymorphism correlations; isogene discovery, breast cancer; vascular lesions
6. Genome Therapeutics (USA)	High-resolution polymorphism database
7. Genostic Pharma (UK)	Polymorphism and allele frequency analysis

Contd....

Company	*Area*
8. Genset (France)	High-density biallelic maps (60,000 markers)
9. Janssen Phannaceutica (Belgium)	Cytochrome variation
10. Millenium (USA)	SNPs in pharmogenomics
11. Mitokor (USA)	Mitochondrial genome
12. Nova Molecular (Canada)	CNS disease genetic profiling
13. Sequana Therapeutics (USA)	High-throughput genotyping
14. Variagenics (USA)	Haplotyping SNPs for use in clinical trials

It has also been shown, that variation in response to most drugs is explained by pharmacokinetics (involving enzymes) rather than by pharmacodynamics (generally involving proteins other than enzymes). Therefore, pharmcokinetic studies have focussed on the variation in drug response and its correlation with variation in cytochrome P450 isoenzymes, dihydropyrimidine dehydrogenase and thiopurine methyltransferase (TPM). Similarly pharmaco-dynamic studies, focussed on cholesterylester transfer protein and serotonin transporter, have also been conducted.

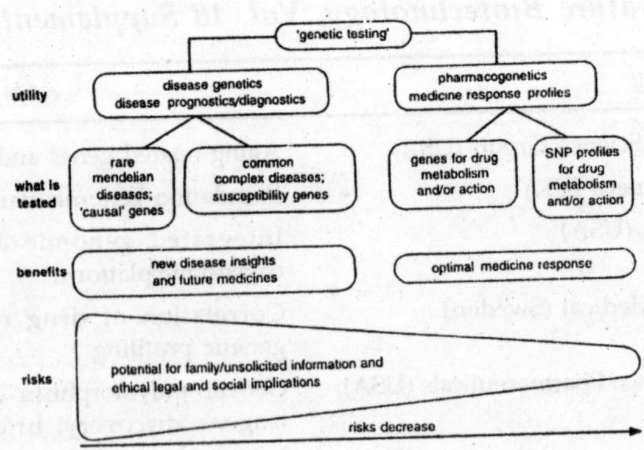

Details of the utility and nature of genetic tests that can be conducted for personalized medicine and the associated benefits and risks with each test (notice that the risk involved decreases as we shift from use of Mendelian genes to SNP profiling).

However, there are also, ethical issues associated with pharmacogenomics, that will be raised and debated in future. Since the patient's confidential genomic information will be stored by the physician, an unwarranted use of pharmacogenomic data will have to be regulated.

Therefore, major benefits and dangers of unregulated developments will have to be compared and it will have to be ensured that the pharmacogenomic information is used only for the benefit of the patient.

10

Gene Therapy

During the last two decades of the 20th century (1980-2000), more than 200 genes responsible for a variety of genetic disorders in human beings were identified. The causes of disorders in the form of specific mutations in these genes have also been elucidated. Therefore, if the known defects in genes are rectified, or if the defective gene in each case replaced by a normal healthy gene, then the corresponding genetic disorder may be corrected. Such a treatment of human genetic disorders are described as gene therapy. However, gene therapy need not always be targeted at either the correction of the defective gene or suppression of its expression or even introduction of a normal healthy gene, but may also be targeted to peripheral or even epigenetic aspects of the pathogenesis pathway.

For instance, while applying gene therapy to cancer, one may not like to correct the oncogene or introduce the cancer suppressor gene, but may only introduce either the genes having anti-tumour function in cells (e.g. tumour-infiltrating lymphocytes or TILs) or introduce cytokine genes into the tumour cells, thus indirectly suppressing tumour growth. In other words, gene therapy involves treatment of genetic disorders by insertion of a gene into the body of an organism, in order to correct the disorder. This sometimes involves, replacement of the defective gene by the normal gene in a functional sense, but not in a physical sense. The healthy gene is introduced into the appropriate cells of the patient, a process that is technically described as gene delivery. Effective systems of gene delivery have been developed in many cases involving retroviruses and a variety of other viral and non-viral vector systems. More recently, nanotechnology has also been utilized for transfer of genes in particles that measure <500nm in diameter.

More important than the above gene correction method is the gene augmentation method, where normal foreign gene sequences for the defective gene are introduced. A number of efficient methods for this purpose are already available, where a number of copies of the desired gene are introduced in the cell and are made to express at high level. Expression and transfer vectors, in the form of a number of viruses are now available to achieve this goal. In view of this, maximum progress in the area of gene therapy has been achieved through the above gene augmentation model involving vector mediated DNA delivery system.

Once the gene correction or gene augmentation has been achieved at the cellular level (in the cells obtained from the affected organ depending upon the disease), the modified cells can be implanted into a suitable region either in an organ of the patient or in the embryo. Direct delivery of the DNA (carried by the vector) into the living cells of the body has also been suggested in several cases, so that both *in vitro* and *in vivo* introduction of corrected gene has been suggested. Therefore, gene therapy can be used at the following two different levels : *(i) embryo therapy*, in which the genetic constitution of embryo at the post-zygotic level is altered, so that the inheritance will also be altered, and *(ii) patient therapy,* in which cells with healthy gene may be introduced in the affected tissue, so that the healthy gene overcomes the defect without affecting the inheritance of the patient. It is believed that in future, gene therapy of both types will be possible.

Research in gene therapy, as a biomedical discipline, started in mid 1960's and the first gene therapy experiment was conducted in 1990 on a human patient {a four - year old child suffering with adenosine deaminase (ADA) deficiency, which had wiped out her immune system, crippling her ability to fight infectious diseases}. During 1990-2000, hundreds of approved clinical trials, involving more than a thousand patients have been conducted in pilot phase I experiments, so that opportunities in gene therapy are attracting interest not only from research organizations in the public sector, but also from both small and large pharmaceutical companies. Different aspects of gene therapy including its possibilities and limitations will be discussed in this chapter.

Human Diseases Targeted for Gene Therapy

Human diseases, that are targeted for gene therapy, may be due to defects in single genes or may be acquired under a variety of circumstances due to genetic predisposition to these diseases. Diseases belonging to both these classes can be treated through gene therapy. Single gene controlled diseases are transferred from generation to generation. The genes for these diseases are generally carried in heterozygous condition in the normal individuals and the disease is caused by the recessive allele in homozygous condition. If the disease is caused by the dominant allele, gene therapy becomes difficult, because, not only a normal allele needs to be introduced, but the defective dominant allele needs to be removed. The heritable Mendelian diseases include phenylketonuria, haemophilia, cystic fibrosis (CF), etc. Acquired diseases include diseases, which are caused mainly due to somatic mutations in some genes, so that they are caused due to specific genes, but seldom transmitted to the germ line. Majority of cancers belong to this class, although heritable cancers are also known now. Acquired diseases are also caused due to genes, which predispose an individual to a disease, like cancer, coronary artery disease, Alzheimer's disease, etc. These diseases are sometimes Mendelian, but are more often caused by a combination of genetic predisposition and environmental factors, or even solely due to environmental factors.

For instance, among heart diseases, a small proportion (e.g. type III hypercholesterolaemia) is caused due to a deletion or mutation either in the gene encoding low density lipoprotein (LDL) receptor, which imports cholesterol into the cell, or in a gene encoding carrier protein apolipoprotein B. But most heart attacks are caused due to a mixture of genetic and environmental causes, which means that there is a predisposition to heart disease, which requires environmental triggers or abuse, to become apparent. Knowledge about such predisposition through genetic testing in such cases may enable the predisposed person to modify the environment. Somatic gene therapy can be used for all the above classes of diseases.

Some Human Diseases Targeted for Gene Therapy.

I. Diseases caused by single gene defects shown in parentheses (targets for gene therapy)

1. Cystic fibrosis (CFTR);
2. Diabetes mellitus;
3. Pituitary dwarfism (hGH);
4. Emphysema (a-1-antitrypsin);
5. Familial hypercholesterolemia (LDL receptor);
6. Haemoglobinopathies - thalassemia major and sickle cell anaemia - (β globin);

7. Haemophilia - A, B (Factor VIII, Factor IX);
8. Lysosomal disease;
9. Gaucher's disesase (glucocerebrosidase);
10. Glycogen strorage diseases;
11. Phenylketonuria (phenylalanine hydroxylase);
12. Severe combined immunodeficiency diseases or SCID (ADA);
13. Duchenne muscular dystrophy (dystrophin);
14. CNS disorders - Leisch - Nyhan syndrome (hypoxanthine phosphoribosyl transferase) and Tay-Sachs disease (hexosaminidase).
15. Immunodeficiency (adenosine deaminase);
16. Hyperammonaemia (ornithine transcarbamylase).

II. Diseases acquired due to genetic predisposition (amenable to gene therapy using gene shown in parentheses)

1. Cancer (interleukins, TK, tumour suppressor);
2. Cardiovascular diseases - Myocardial infarcts or MI (tPA); prevention of blood clots (tPA), hypercholesterolaemia (LDL receptor)
3. Neurodegenerative diseases - Alzheimer's disease (NGF), Parkinson's disease (TH);
4. Joint disorders - Rheumatoid arthritis (cytokine, IL-1 antagonists);
5. Infectious disorders - AIDS (HIV antigens, cytokines, TK genes, cytomegalovirus or CMV infections).

CFTR = Cystic fibrosis transmembrane conductance regulator; hGH = human growth hormone; LDL = low density lipoprotein; ADA = adenine deaminase; TK = thymidine kinase; tPA = tissue plasminogen activator; NGF = nerve growth factor; TH = tyrosina hydroxylase; IL-1 = interleukin-1

Vectors and other Delivery Systems for Gene Therapy : Gene delivery for gene therapy, is achieved by one of the following two methods :

(i) viral transduction, involving introduction of a modified virus carrying foreign gene and

(ii) physical transfection involving the use of either 'chemical methods' like calcium phosphate, liposomes and molecular conjugates or 'physical methods' like electro-poration and particle acceleration (biolistics).

Since transduction gives stable integration and autonomous replication, vector mediated gene transfer is preferred and a variety of vectors have been tried (Table). However, one of the serious problems is the nonavailability of an ideal vector, which must achieve regulated and sustained expression of foreign genes in specific cells or tissues. This requires that a foreign gene be included in a gene construct, which either replicates autonomously or gets integrated with an active region of the genome of the host.

Furthermore, the whole process must be safe, efficient and selective. Among the available vectors, each vector satisfies some of these features but none fulfils all the requirements.

Comparison of Gene Delivery Systems for Gene Therapy

Delivery system	*Integration*	*Transduction*	*Comments*
I. Viral vectors			
1. Retrovirus	+	–	Low titre
2. Adenovirus (AV)	+	—	
3. Adeno-associated virus (AAV)	+	–	Helper virus needed

Contd...

Delivery system	*Integration*	*Transduction*	*Comments*
4. Lentiviruses (Herpes simplex virus)	+	?	Pathogenic
II. Physical methods			
1. Receptor mediated	+	+	Enhanced by
2. Direct injection	–	+	Muscle
3. Lipofection	+	+	Gene gum

Viruses as Vectors

Retrovirus (RV) as a vector. Among the vectors, the most widely used vector is retrovirus, because it can introduce genes into a single active region of the chromatin, thus making it a permanent part of the genome for long term expression.

However, there are a number of problems associated with the use of retroviruses as vectors, but solutions to some of these problems have been found. Significant improvements have actually been made in retroviral vectors through incorporation of synthetic and heterologous components.

Adenovirus (AV) as a vector. Adenovirus is a common virus that infects the human respiratory tract and eyes. It can be easily grown in laboratory and infects both dividing and non-dividing cells. It expresses genes without inserting them into the host's genome, so that although it poses no risk of cancer, but it stimulates strong immune response, thus clearing the vector from the body and making it ineffective for long-term therapy. High doses may be required to transfer enough genes for health benefit and this may cause toxic effects.

Present Day Gene Therapy Vectors and their Characteristic Features.

Features	*Viral vectors*					*Non-viral vectors*		
	RV	AV	HV	PV	AAV	Mol. conj	Transfection	Liopsomes
Insert capacity (kb)	8	7-8	720	725	4.5	—	—	—
Integrates	Yes	No	No	No	No	No	No/rare	No
Oncogenic	Yes	Yes	?	?	?	?	?	?
Viral proteins expressed	No	Yes	Yes	Yes	Yes	No	No	No
In vivo delivery	poor	Yes	Yes	Yes	Yes	?	Yes	Yes

RV = retrovirus; AV = adenovirus; HV = herpes virus; PV= AAV = adeno-associated virus

Characteristics of Different Types of Vectors Used for Delivery of Genes

Retrovirus	*Adenovirus*	*Other viral vectors*	*Non-viral vectors*
RNA genome (e.g.MMLV) associated virus or AAV, HSV1)	Linear double stranded DNA genome complex	DNA (adeno-	DNA-protein
Reverse transcribed	Does not integrate in host genome; slowly lost	It integrates	DNA entrapped in endosomes
Viral genome integration in host genome; is not lost	Human kidney cell line supplying EIA products-dividing cells not necessary		Cultured hepatocytes
Suitable host : proliferating tissue	EIA region of virus, needed for gene expession and replication, is deleted and replication by therapeutic gene		
Therapeutic gene is inserted in viral genome	Can infect non-dividing cells (major advantage)		

Major Problems with Retroviral (MOMLV) Vector and Possible Solutions

Problem	*Solution*
1. Regeneration of replication competent retrovirus during vector production	1. Replace murine cells with human helper cells; replace MLV vector with non-homologous vectors
2. Insert size limited to ~ 8 kb	2. Use synthetic particles or MAS vectors
3. Rapid inactivation *in vitro*	3. Modify helper cells, *gag* and *env* genes; use synthetic particles/conjugate/liposomes
4. Infect only dividing cells	4. Add HIV matrix protein gene to helper systems

Adeno-associated Virus (AAV) as a Vector : AAV is a parvovirus, which readily infects dividing and non-dividing cells. It needs a 'helper' adenovirus to replicate, and integrates into host DNA at a safe location causing no risk for cancer. However, it is difficult to grow to high concentration (relative to adenovirus) and has a small genome thus restricting the amount of therapeutic DNA, although techniques are being developed to increase AAV's gene carrying capacity. It has been used to transfer gene for clotting factor IX into patients suffering with hemophilia B.

Lentiviruses as a Vector : HIV and Herpes Simplex Virus (HSV) are lentiviruses, which can introduce genes into dividing and non-dividing cells of nervous system, which are resistant to other vectors.

Common Problems with Most Viral Vectors : Some of the common problems with majority of viral vectors include the following : (i) a limited DNA capacity, (ii) expression of viral genes, (iii) initiation of antiviral immune response, (iv) reversion to a replication competent state and (v) decreasing expression over time. In view of this, in the year 1995, National Institute of Health (NIH), also established three 'national gene vector laboratories', so that high quality gene transfer agents may become available for gene therapy. Each centre will plan to focus on specific vectors gene delivery systems including the following : (i) retroviruses, adeno-associated virus, (ii) liposomes, naked DNA and DNA-coated pellets and (iii) adenovirus and other DNA viruses.

Non-viral DNA Delivery Systems

It is known that about 75% of recent clinical protocols involving gene therapy make use of viral-based vectors for DNA delivery. However, in clinical trials, none of these viral vectors could be shown to be safe. Therefore, non-viral DNA delivery systems have been developed both for gene therapy and DNA vaccination for treating and controlling diseases.

Most of these available DNA delivery systems are though versatile and safe, they are less efficient than the virus-mediated DNA delivery systems discussed above. The low efficiency of non-viral DNA delivery is due to the following three major barriers to DNA delivery : (i)low uptake across plasma membrane, (ii) inadequate release of DNA molecules and (iii) lack of nuclear targeting. These three barriers have been addressed in the efforts to develop efficient DNA delivery systems, which can be broadly classified into (i) mechanical/ electrical methods and (ii) chemical methods.

Mechanical/electrical Methods : Direct injection of naked DNA into the cell is simple and therefore appealing, although there are problems associated with it, since only single cells can be manipulated, and the process ill slow and laborious. The three methods that have been tried include : (i) low voltage electroporation or microinjection; (ii) particle bombardment and (iii) pressure-mediated DNA delivery. The details of electroporation and particle bombardment will be discussed later, and are commonly used with cultured cells except where limited local expression of delivered DNA is adequate to achieve immune response (e.g., DNA vaccination in epidermal or muscle cells). Pressure-mediated delivery gave 50% efficiency of delivery, when tried with cardio-vascular tissue. Hydrodynamic force has also been used for DNA delivery to hepatocytes with 40% efficiency. Ultrasonic nebulization has also been tried.

Chemical Methods : The use of a chemical to enhance DNA uptake by the cell is considered easy and versatile. The techniques can be broadly classified into those based on : (i) 2-diethylamino ether (DEAE)-dextran, (ii) calcium phosphate, (iii) artificial lipids, (iv) proteins, (v) dendrimers, etc. In all these techniques, DNA is complexes with a chemical and is deposited onto the cells, so that they are internalized by endocytosis and the DNA is released at the desired target site. The major limitation of these methods is toxicity upon systemic administration.

(a) *Efficient Uptake of DNA During a Pulse of DNA Complexes with a Chemical* : During 1965-75, DEAE-dextran-DNA and calcium phosphate-DNA complexes were used for internalization by the cells through endocytosis. These methods can be largely used for *in vitro* transfection and suffer with problems of unstable and variable transfection. During 1970-1990, a variety of artificial lipids-based DNA delivery systems were developed, which are the most commonly used methods today. Lipofectin-DNA complexes were the first chemical systems that could be used in animals and liposome-mediated DNA delivery has been used in humans. The cationic peptide poly-L lysine (PLL), polyamidoamine (PAMAM) dendrimers and polyethyleneimine (PEI) are other chemicals, which vary in efficiency (20%-80%), but are toxic.

(b) *DNA complexes with polymers for controlled release on a long-term basis.* The above chemical methods involve exposure of cells to a pulse of DNA complexes with a chemical, so that a short period is available for uptake of DNA. Therefore, biocompatible polymers have been used, which allow controlled release of DNA on a long-term basis without repeat administration. For instance, biodegradable poly (D) L-lactide-co-glycolide (PLG), microparticles/microspheres and poly ethylene-co-vinyl acetate (EVAc) matrices have been used for long-term controlled release of naked DNA within the cell.

Synthetic Particles as Vectors

Synthetic particles may combine useful attributes of different classes of vectors and avoid viral pathogenicity. These synthetic vectors include liposomes or molecular conjugate vectors or a combination of both. These synthetic vectors can accommodate genes, more than 8-kb long and can be engineered for entry into the cell through receptor mediated endocytosis

(RME). They are so engineered that after entry into the cell, they leave endosome and enter the nucleus and not the lysosome. For this purpose, a receptor binding moiety (e.g. asialo-orosomucoid, which is a ligand for the asialoglycoprotein receptor of hepatocytes) and a DNA binding moiety are incorporated. Microtubule inhibitors may block uptake by lysosomes, and nuclear localization signals (NLS) may facilitate entry into the nucleus. In contrast to vector-mediated gene transfer, transfection is though safe and efficient but involves no mechanism for integration or autonomous replication. It allows only transient expression. In view of this, efforts are also being made to combine transduction and transfection for efficient and safe gene transfer providing long-term expression.

Selectivity and Long-term Expression

The delivery system and the gene used for gene therapy should also provide selectivity at the following levels : (i) it should enter non-germline cells; (ii) it should selectively integrate at a specific site within a genome and (iii) it should be selectively expressed (over time and space). Even if this selectivity is achieved, long-term expression may still remain a problem. Stable integration may partly solve the problem and may be achieved through the use of retroviral reverse transcriptase, retroviral integrase, transposase and parvovirus 'rep' proteins. The enzymes may be used with the vector, or else DNA encoding these enzymes may be included in non-integrable form for transient expression to facilitate stable integration. Autonomous replicating units like mammalian artificial chromosomes (MACs) or autonomous replicating circular minichromosomes (found in some cancer cell lines) may also be used for achieving long-term expression.

Transient Expression for Gene Therapy

Depending on the disease being treated through gene therapy, expression of the introduced gene may be required throughout the body or its expression may be restricted to particular tissues. Also, although in majority of cases of gene therapy, sustained genetic modification is needed, in some cases, a transient expression of therapeutic gene may be desirable. Two such studies involving central nervous system were conducted and published in September 2000 (Nature Biotechnology). In one of these studies, a viral vector was used, which carried a gene, whose transient expression was triggered by the neurological insult itself. In the other study, poliovirus replicon vector was rendered non-infectious and was used to carry the desired gene for transient expression in transgenic mice; the latter carried human poliovirus receptor (PVR). In the first study the gene expression could be modulated by pathophysiological variations due to the stroke itself, while in the second study the vector itself limited the replication thus limiting the transgene expression. This transient expression of desired gene in the affected brain area provides neuroprotection from a stroke.

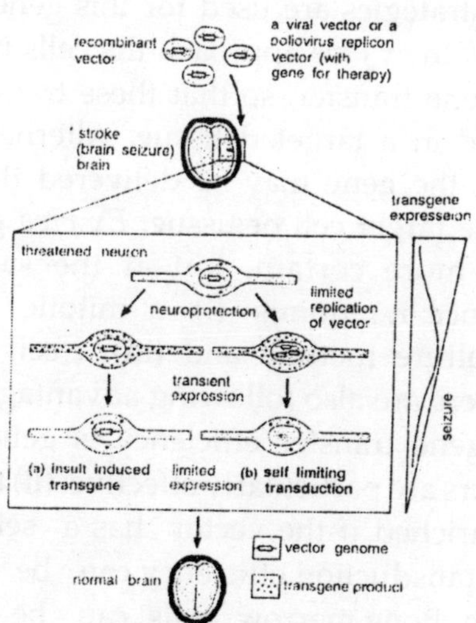

Use of poleovirus replicon vector (it was modified, so that it could replicate, but is non-infectious) for transient gene expression of a therapeutic gene that can be used during brain seizure (redrawn from Nature Biotechnology, September, 2000).

Splicesome Mediated trans-RNA splicing (SMART) : A novel approach for gene therapy, described as SMART, was reported in January 2002. In this approach, an exogenous DNA, coding for trans-splice mRNA (mRNA, which will undergo trans-splicing with native endogenous mRNA) is introduced as a part of gene therapy. The exogenous DNA is designed to contain the following sequences in 5' to 3' order : (i) a binding domain (at 5' end) that hybridizes to the intron present towards 5' to the defective exon, to be corrected, (ii) a spacer and (iii) at its 3' end, a 3' splice branchpoint and acceptor, which lies 5' to the defective exon. The mRNA transcribed from this introduced exogenous DNA and the defective endogenous mRNA are spliced and rejoined in such a manner, that the native mRNA will be corrected for the mutation.

Ex Vivo vs in Vivo Gene Therapy

In gene therapy, recombinant vector with the therapeutic gene is used for gene delivery to specified cells and tissues. Two different strategies are used for this gene delivery, i.e. *a vivo* and *in vivo*. In *ex vivo* approach the cells may be cultured and used for gene transfer, so that these transfected cells are then introduced in a targeted tissue. Alternatively in the *in* vivo approach, the gene may be delivered through a vector directly into the target cell or tissue. *Ex vivo* gene delivery is common and more certain, but at the same time more problematic, since: it requires (i) a mitotic cell population; (ii) a tissue culture method and (iii) a cell transplantation technology. There are also following advantages of an *ex vivo* approach, (i) gene transfer efficiency is generally high and retroviral vectors are particularly effective: (ii) the transduced cells can be enriched if the vector has a selectable marker gene and (iii) transduction efficiency can be assessed before re-implantation. Bone marrow cells, can be multiplied and modified and are the common candidates for *ex vivo* gene delivery, but for other types of cells, *ex vivo* modification may be difficult. However, the use of embryonic stem (ES) cells may be really useful. For this purpose, although earlier ES cells were available only from mouse, rat, pig, rabbit, etc. but human ES cells have also become available recently (1998-2002).

Despite some advantage of *ex vivo* gene delivery, *in vivo* gene delivery would be preferred, if feasible. However, one major precaution required during *in vivo* gene delivery is that the gene be delivered only to the targeted cells and no other cells, and in no case to the germ line cells. Retrovirus delivers the gene only to dividing cells and therefore is safe for delivery of gene to cancer cells. But in other cases receptor mediated endocytosis (RME) may have to be used, so that the vector will have to carry a ligand for the receptor available on the target cells. Sometimes, instead of targeting specific cell types, selective expression of a transferred gene can also be achieved by the use of tissue specific enhancers/promoters. However, the disadvantages of *in vivo* gene delivery include the following (i) specificity and low efficiency of stable gene transfer, (ii) for clinically useful therapeutic application, repeated treatments may be needed raising the problem of a host immune response.

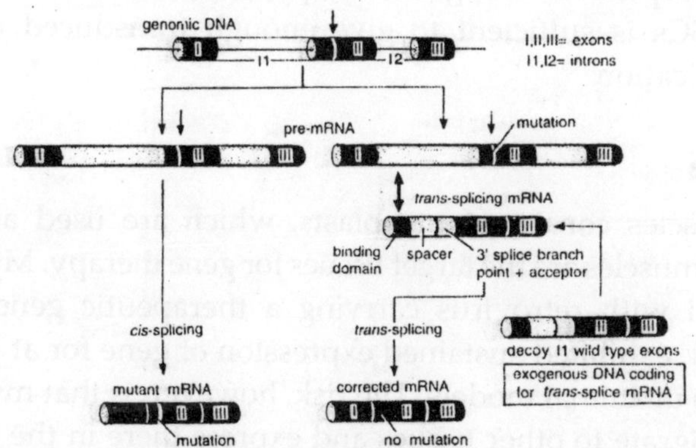

Spliceosome-mediated trans-RNA splicing (SMART) to correct a hereditary disorder at the RNA level; notice that a mutation occurs in exon II, which gives a mutant mRNA due to cis-splicing (shown on the left), and gives a corrected mRNA due to trans-splicing if a DNA sequence carrying the wild type exons is introduced (shown on the right), (redrawn from Nature Biotechnology, January, 2002).

In view of the above advantages and disadvantages of *ex vivo* and *in vivo* approaches, the choice will depend on the relative ease of *in vitro* culture *(ex vivo* approach) and the gene transfer efficiency of the target tissue *(in vivo* approach).

Target Tissue of Choice for Gen-Delivery System : Both in *ex vivo* and *in vivo* approaches, there will be a target tissue in the body, where either the cells transfected in culture, will be transferred, or the vector carrying the therapeutic gene will be targeted, either for integration in the DNA of target cells (retrovirus) or for transient expression (adenovirus). Following are some of the target tissues.

Bone Marrow

Bone marrow is a hematopoietic system, which is a suitable target for gene therapy, since well developed procedures for bone-marrow transplantation are available. The pluripotent hematopoietic stem cells (HSC) constitute .01 to 0.1% of human bone marrow cells which are, therefore, considered as an ideal target tissue. Gene transfer to HSC is also considered to be the best example of *ex vivo* gene therapy, and transduction of even few HSCs is sufficient to give enough transduced cells by multiplication.

Muscle

Muscles consist of myoblasts, which are used as target cells, if muscles are the target tissues for gene therapy. Myoblasts infected with retrovirus carrying a therapeutic gene, when injected, exhibited sustained expression of gene for at least six months in animal models. The risk, however, is that myoblasts may migrate to other tissues and express there in the absence of tissue specific expression.

Adenovirus vector containing a reporter gene has shown efficient transduction of muscle after direct injection in mouse. Myoblasts are also suitable for direct gene transfer using *in vivo* approach. Stable gene expression of injected DNA (present as episomes) was achieved in cardiac and skeletal muscles of mouse, although the efficiency was low in primates, when 'gene gun' was tried.

Liver

Liver contains hepatocytes, which can be used as target cells, using both *ex vivo* and *in vivo* approaches. However, they can not be extensively manipulated in culture, because they undergo few cell divisions. Further, the transduction efficiencies with retroviral vectors are rather low. Nevertheless, retrovirus-mediated *ex vivo* transfer of disease related human genes has been successfully achieved Hepatocytes, which are generally non-dividing cells, can be activated into cell division by hepatactomy, so that retrovirus mediated *in vivo* gene transfer can be achieved. Adenovirus, adeno-associated virus and some non-viral vectors have also been used for *in vivo* targeting of DNA to the liver.

Other Tissues (Brain, Trachea) : Cells like brain cells or neurons (neurological disorders) and those belonging to tracheal epithelium or lung (e.g. 'cystic fibrosis' diseases) are difficult targets, because cells can not be easily harvested for *ex vivo* approach and the tissues are generally inaccessible for *in vivo* approach. However, HSV-1 vectors have been used to transduce neurons both *ex vivo* and *in vivo*. Similarly adenovirus mediated and liposome mediated gene delivery to respiratory epithelium has been achieved in some animal models (e.g. rat). Cell types including keratinocytes, fibroblasts, myoblasts, vascular endothelial cells/smooth muscle cells are also used for delivering genes whose products are secreted into the circulating blood. *In vivo* gene transfer into a specific blood vessel and lymphocytes has also been achieved.

In-Utero Gene Therapy or IUGT (Before Symptoms Appear) : There are about 400 approved clinical gene therapy protocols available worldwide, but all of them treat paediatric and adult patients after birth. However, genetic diseases can also be treated by the gene transfer in the fetus and the technique is described as 'in-utero gene therapy (IUGT)'. For a number of neurological genetic diseases (e.g. Tay-Sachs, Lesch-Nyhan, etc.), where irreversible damage may be caused during gestation before the birth, treatment may be needed before birth to allow the birth of a normal baby. However, decision about the target tissue and the availability of technique to safely and efficiently deliver the gene at the target tissue (e.g. brain) are the major limitations for treating these diseases at the level of fetus. In view of this, diseases involving an easily accessible cell type (e.g. haematopoietic stem or progenitor cells or HSCs) will be the choice for IUGT. These diseases may improve immuno-deficiency, thalassemia, osteopetrosis, etc.

Since IUGT should lead to correction for life-time, the gene should integrate in patient's genome, for which only the retroviral vectors are suitable. For this purpose, HSCs may be genetically engineered *ex vivo* and transplanted back into the patient. Some success in this approach has been achieved in animal systems suggesting that it is a viable approach. Several safety and ethical issues need to be dealt with before clinical trials using IUGT will be permitted.

Gene Therapy for Genetic (Heritable) Diseases

Neurological and Other Metabolic Disorders : Central nervous system (CNS) is a difficult target for gene therapy, but efforts are being made to use gene therapy for a variety of CNS defects whose genetic basis is known now (Table). However, the following aspects of neurological disorders make gene

Gene therapy models for some neurologual disorders

Disorder	*Transgene*	*Target tissue/orgain*
I Genetic disorders		
1. Phenylketonuria	Phenylalanine hydroxylase	Hepatocytes or direct dilivery to liver
2. Leisch-Nyhan disease	HPRT	Surrogate cells or direct delivery to brain

Contd...

Disorder	*Transgene*	*Target tissue/orgain*
3. Lysosomal storage disorder	Lysosomal exumed	Bone-marrow cells or delivery to brain
II Neurodegeneratine disorders		
4. Alzheimer's diseose	NGF or other trophic factors Tyrosine hydroxylase	Surrogate cells or direct delivery to basal forebrain.
5. Parkinson's disease		Surrogate cells or direct delivery to basal ganglia or mid brain

therapy difficult for these diseases : (i) compart-mentalization of neural systems, (ii) complexity of neuronal interactions, (iii) poor accessibility of CNS due to physiological blood-brain barrier, making introduction of gene-transfer vectors difficult. Despite these problems, at least some neurological disorders of central nervous system (CNS) and peripheral nervous system (PNS) are becoming amenable to gene therapy. Both *ex vivo* and *in vivo* approaches are being tried for diseases like the following : (i) Huntington's disease (HD), (ii) brain tumours, (iii) familial Alzheimer's disease (FAD), (iv) familial amyotrophic lateral sclerosis (ALS), (v) Duchenne muscular dystrophy (MSD).

Ex Vivo Approaches to Gene Therapy for Neurological Disorders : Although success in treating neurological disorders through gene therapy may take time, it seems likely that a graft of genetically modified cells (which can synthesize and secrete trophic factors or provide components of neurotransmitter pathway) might be effective at least in some cases of neurological disorders, particularly those with focal dysfunction. Parkinson's and Alzheimer's diseases are candidates for this kind of gene therapy (the genetic basis of these diseases, if any, is not fully understood). Some success in this connection has been achieved using rat models. For instance, in Parkinson's diseases, the lack of ability to synthesize the neurotransmitter, dopamine is believed to be the main cause of the disease, so that the grafts of dopamine producing fetal cells or genetically modified cells capable of synthesizing dopamine have been shown to help in correcting the disease symptoms. Similarly, the supply of neurotrophic growth factor (NGF) from hippocampus plays an important role in survival and function of 'cholinergic neurons', which are essential for memory acquisition and retention. Degeneration of these cells in the absence of NGF causes Alzheimer's disease. It has been shown that the supply of NGF or genetically modified cells capable of synthesizing NGF will help correcting the symptoms of Alzheimer's disease. Genetically modified bone marrow or the precursor CNS cells that are capable of differentiating into neurons or glial cells, can also be introduced to deal with some neurological disorders.

In Vivo Approaches to Gene Therapy for Neurological Disorders : DNA mediated gene transfer can be achieved by injection of naked plasmid DNA into the brain although the efficiency of transfer is low, so that more efficient viral vectors may need to be used in future.

Gene Therapy for Cystic Fibrosis (CF) Lung Disease : Cystic fibrosis involves failure of airway defense against bacterial infection and is attributed to loss of function of cystic fibrosis transmembrane conductance regulator (CFTR) protein due to a mutation in CFTR gene. CFTR regulates certain ion channels. The gene *CFTR* carried by a vector like adenovirus has been used both *ex vivo* and *in vivo* gene delivery. Liposomes have also been tried.

Gene Therapy for Acquired Diseases

Gene Therapy for Infectious Diseases : Two different strategies for the treatment of infectious diseases are being explored. Both strategies involve the use of genetically manipulated cells. (i) The first strategy termed 'intracellular immunization' is designed to render cells resistant to viral replication, thus limiting the spread of virus. (ii) The second strategy termed 'genetically engineered vaccines' or 'DNA vaccines', makes use of genetically modified cells expressing viral gene products to enhance antiviral cellular immune responses.

Intracellular Immunization (Rendering Cells Resistant to Viral Replication / Gene Expression) : In the year 1988, Baltimore coined the term 'intracellular immunization' for genetic modification of target cells (prone to viral infection), making them resistant to intracellular pathogens. However, 'intracellular resistance', 'intracellular interference' or 'intracellular inhibition' could be better terms. AIDS is an example of the approach, because insertion of resistance genes into hematopoietic cells (that are prone to infection by HIV) or the use of gene based vaccines could reduce, if not eliminate, the spread of virus. The resistance may be achieved through (i) antisense RNA, (ii) ribozymes, (iii) RNA decoys (exploiting steps of HIV replication), (iv) use of altered HIV gene products having transdominant (TD) mutant phenotype (e.g. altered forms of Gog, Rev, Tat and Env).

Genetically Engineered Vaccines (Also called DNA Vaccines) : The approach involving genetically engineered vaccines, can induce production of either the antibodies or the cytotoxic T cell receptors. Introduction of genes that encode T cell receptors has been considered a better alternative. The above resistance genes or the gene vaccines need to be delivered to hematopoietic target cells of the patient and this may be achieved using either the retrovirus based gene transfer systems or the vectors based on adeno-associated virus (AAV). The advantages and disadvantages of using these vectors have been discussed elsewhere in this chapter.

The efficacy and safety of the use of resistance genes or the genes encoding T cell receptors for gene therapy have been tested in immortal cell lines and in animal models, but both these systems have their limitations that need to be dealt with, before clinical trials on human subjects can be conducted.

Gene Therapy for Cardiovascular Diseases : Several cardiovascular diseases are due to abnormalities in muscles of the hearts or blood vessels. In cases of arteries blockage, when arteries are opened using baloon angioplasty, reblockage (also called 'restenosis') occurs in 50% cases and can be treated by 'gene therapy'. Gene therapy has been attempted for several of these diseases. For instance, blockage of arteries caused by proliferation of smooth muscles lining the artery walls, can be treated by gene encoding thymidine kinase (tk) carried by adenovirus. Thymidine kinase makes the proliferating cells susceptible to a drug called ganciclovir. Similarly, the gene for vegf (vascular endothelial growth factor) has been used for growth of new blood vessels for the heart. The vegf gene may be transferred through a virus or via liposomes.

Gene Therapy for Cancer (Somatic Mutations) : It has been shown that cancer in most cases has a genetic basis, but in majority of cases it is a disease, which is acquired during the life span of an individual. Such an acquired cancer results due to somatic mutations, although it can also be inherited from the parents.. In either case, gene therapy can be used to selectively target and destroy cells, by using one of the following approaches :

(i) Whenever, cancer is caused, either due to loss of tumour suppressor gene like p53 gene or due to mutation in this gene, insertion of a copy of wild type p53 gene has been shown to result in the death of tumour cells.

(ii) If cancer is caused by overexpression of an oncogene like *K-RAS*, an antisense gene has been shown to disable overexpression of this gene.

(iii) Therapy using immunostimulatory genes, such as interleukin (IL-2) gene may also lead to immunological destruction of tumour cells.

(iv) Insertion of an antisense gene that stops production of insulin like growth factor (IGF-1) in some tumours also leads to destruction of tumours, since these tumours produce high level of IGF-1.

In all the above approaches, the most important problem is the need for a more efficient gene delivery system. No gene delivery system is 100% efficient, unless germ-line therapy is used. However, non-integrating vectors such as those derived from adenovirus have been found to be efficient. In contrast, if integrating vectors can be developed, advance delivery of genes in susceptible patients may prevent development of malignancies.

Clinical trials for untested technologies are also appropriate ethically for those cancers which are incurable, and when life expectancy is a few weeks/months, rather than years as in other genetic diseases. In such cases approval of trials is also easy.

Ex Vivo Gene Transfer in Tumour Cells, Fibroblasts or T Lymphocytes : In this approach tumour cells from the patient are grown in culture, immunostimulatory genes are inserted *in vitro* and the genetically engineered tumour cells are reinjected into the patient. This not only destroys the tumours, but also vaccinates the patient against the recurrence of the tumour. In most such cases retroviral vectors are used for delivery of either the genes encoding cytokines (e.g. IL-2, IL-4, IL-6, IL-β \ TNF-α , etc.), or the antisense genes blocking production of growth factors. In some cases, the gene is inserted into patient's fibroblasts, which are mixed with irradiated tumour cells from the patient and reinjected. This is preferred because culturing fibroblasts is easier than culturing tumour cells from a large number of individuals. Tumour infiltrating T lymphocytes (TILs) have also been used for genetic engineering to deliver cytokines directly onto the tumours, although their transformation is difficult and they down-regulate expression of cytokine gene.

Sensitivity genes have also been used, which increase sensitivity of the tumour cells to certain drugs that will thus become toxic and kill tumour cells. One such gene is thymidine kinase gene of the herpes simplex virus *(HSVtk).* It imparts to the cancer cells sensitivity to anti-herpes drug 'ganiclovir (GCV)', which is phosphorylated due to HSVtk and inhibits DNA polymerase, causing cell death. The approach has been used in certain brain cancers and ovarian cancers.

Ex-vivo Gene Transfer in Bone Marrow Stem Cells : Gene therapy in cancer patients is also used for protecting hematopoietic stem cells from the toxic effects of chemotherapy. Gene conferring multidrug resistance (MDR-1) has been used for this purpose. Bone marrow cells with MDR-1, when injected, have been shown to protect stem cells *in vivo.*

In Situ Genetic Alteration of Cancer Cells : Both liposome mediated and retrovirus mediated gene transfer has been attempted to modify tumour cells *in situ.* Genes for antigens like HLA-B7 and β2-microglobulin have been transferred through liposomes and expressed on cell surface inducing `immune reaction against tumour cells. Fibroblasts, that produce retrovirus vector (also called retroviral vector producer cells or VPCs) have also been directly injected into growing brain tumours, for retrovirus mediated transfer of HSVtk gene. In animal models, and human patients, this led to complete tumour destruction in 80% of animals. *In vivo* retrovirus mediated transfer of p53 tumour suppressor gene, or antisense K-RAS gene or γ-interferon gene have also been tried for some kind of lung cancers in humans.

Nanotechnology for Drug Targeting and Gene Therapy : Selective delivery of drugs to specific areas in the body is necessary to maximize drug action and minimize side effects. The 'magic-bullet concept' represents an early description of drug targeting paradigm. Such targeting can be achieved at the organ level, cell level or even at the intracellular level and the drugs may involve prodrugs (which release a drug at the targeted site), monoclonal antibodies and polymeric and colloidal carriers.

Nanotechnology involves the use of nanoparticles (<lum, usually <500nm in diameter) used as particulate carrier systems for drug delivery. These particulates include not only solid structures in the form of nanocapsules, but also includes liposomes and emulsions, that can reach specific tissues or even specific sites within a cell. They can be used for delivery of drugs, vaccine administration and for diagnostic imaging.

Designing of Targeting System, Keeping in View the Fate of Nanoparticles after Intravenous Administration : A targeting system needs to be appropriately designed keeping in view the fate of nanoparticles after intravenous administration. After the particles are injected into the blood stream, they are conditioned or coated by elements like plasma proteins and glycoproteins available in the circulatory system. This process is called 'opsonization', which renders the particles recognizable by the major defense system of the body. The macrophage cells (called 'Kupffer cells') of the liver and other macrophages remove these opsonized particles. Particles having hydrophobic surface are more efficiently coated, but those having hydrophilic surface resist this coating and are therefore cleared more slowly. Therefore, nanoparticles having hydrophilic surfaces, should be preferred for gene therapy.

Passive targeting of nanoparticles. Although uptake of nanoparticles by Kupffer cells may be a hindrance in delivery of drug to the target, at least in some cases, this feature can be exploited to help treatment. This is true in cases, where the macrophages themselves are directly involved in the disease process, as in case of leishmaniasis, listeria and candasis. In other cases, capture and degradation of nanoparticles by macrophages can also be exploited to achieve the controlled delivery of drugs to the blood circulation. This is because through macrophages, the particles reach endosomes and then to lysosomes, where they are degraded, so that the drug is released and diffused into the circulatory system. This strategy can be applied in the delivery of anticancer agent adriamycin, and antifungal agent amphotericin.

In cases, where capture by Kupffer cells is a barrier to the targeting of drug, the particles may be disguised, so that they are no longer recognized by Kupffer cells and therefore escape capture by them. The nanoparticles can be disguised partly by making them hydrophilic and partly by stabilizing them using adsorbed polymers. Nondegradable polystyrene, polyoxy-ethylene or a block copolymer are some such adsorbed polymers, which do not allow recognition of liposomes by macrophages and thus prevent opsonization so that the nanoparticles persist within the circulation. Such 'stealth liposomes' are undergoing clinical trials for tumour targeting.

Active Targeting of Colloidal Carriers : Targeting ligands (e.g. monoclonal antibodies, sugar residues, etc.) can also be attached to nanoparticles to achieve active targeting of these particles. The hepatocytes of liver can be an important target in case of hepatitis and also in other cases of gene therapy, when administered gene needs to be expressed in these cells of the liver. The nanoparticles having ligands for active targeting, should not only escape captive by kupffer cells of the liver, but also need to be small enough (>50nm or even >20nm in size) to be able to enter the hepatocytes with the help of specific receptors on these surfaces and through ferestrations that are 100-150nm in diameter.

Gene Therapy using Nanotechnology : Pilot phase I experiments involving gene therapy in human subjects have already been conducted for diseases like cystic fibrosis and muscular dystrophy. Some of the attractive targets for gene therapy include the following : (i) epithelial surfaces of the lungs and gastrointestinal tract; (ii) endothelial cells lining the blood vessels, muscle myoblasts and skin fibroblasts, and (iii) tumour cells. In all these cases, the plasmid DNA carrying the gene sequence should reach the nucleus of the target cell, overcoming all the hurdles (capture by endosome and lysosome) on its way to the nucleus. In the nucleus, it will stay as an episome and express itself.

Some of the Non-viral Vectors for Gene Therapy Based on Nanoparticles

System	*Animal*	*Reute*	*Fate*
1. Naked DNA	Mouse	Intravenous	Liver (Kupffer cells)
2. Cationic lipid	Mouse, men	Aerosol, nasal spray	Lung deposition
3. Polylysine conjugate	Mouse, men	Intravenous	Liver parenchyma
4. Liposome	Rat	Intravenous	Liver (Kupffer cells)
5. Polyvenyl derivative	Rat	Intramuscular	Muscle expression
6. Polycyanoacrylate	—	Cell culture	—
7. Cationic polymers	—	Cell culture	—

Three main types of gene-delivery systems have been described : (i) viral vectors; (ii) non-viral vectors (particles and polymers), and (iii) 'gene guns' for direct injection of the genetic material into the target tissue. Since viral vectors pose some serious problems, the non-viral vectors are the gene-delivery system of choice by many. In this non-viral vector system, negatively charged plasmid DNA is condensed into a non-particulate structure, 50-200nm in size. The use of cationic lipids and cationic polymers gives a compact structure due to interaction between cationic material and anionic DNA. These compact structures also provide increased stability and uptake by the target cells. Some of the Non-viral vectors for gene therapy based on nanoparticles are listed in Table. The targets for gene therapy using nanoparticles include liver hepatocytes, endothelial cells, spleen and lymph nodes, where some success has already been achieved.

Some Initial Success in Gene Therapy **:** In the past, for approved human genetic engineering (or gene therapy), experiments were conducted and partial success was achieved. These experiments include the following :

(i) Neo R/TIL gene marking. In 1989, using Neo R/TIL protocol, the first gene-marked immune cells (tumour-infiltrating lymphocytes =TIL) \yere successfully transferred into patients with advanced cancer. The TILs are those lymphocytes that are isolated directly from tumour and are then grown to large numbers in tissue culture in the presence of cell growth factor (interleukin-2 = IL-2). In Neo R/TIL protocol, an aliquot of cells from TIL is taken; a marker gene is transferred to these cells (the marker gene is neomycin resistance gene = Neo R) using retroviral vector; the marked cell and unmarked cells are grown together and then transferred back to the patient. This is an example of a successful delivery of a marker gene. Thus demonstrating the feasibility of gene delivery approach.

(ii) ADA gene therapy. In, 1900 the first trial of actual gene therapy was conducted in USA. A little (4 years old) girl suffering with adenosine deaminase deficiency (ADA), a lethal disorder, was transfected with lymphocytes bearing the ADA gene carried by retroviral vector. In 1991 another girl (9 years old) began treatment in USA and received more than a dozen transfusions by 1992. Both these patients were doing well. In Italy also, a five year old boy was given a mixture of ADA gene - corrected T cells and bone marrow cells. In Netherlands also bone marrow gene therapy protocol for ADA deficiency was approved in 1992. In 1993, Donald Kohn in Los Angeles separated CD34 cells from umbilical cord blood and transformed them with ADA gene and reintroduced them in the 3 days old children (three). The experiment was partially successful.

(iii) Cancer gene therapy. In 1991 two protocols for cancer gene therapy were initiated in USA. Tumour necrosis factor (TNF, an anticancer agent) gene or *IL-2* (interleukin-2) gene was inserted in TIL tumours cells isolated from the patient and grown in culture. These gene-corrected cells were then injected into the body of patient. Other cancer gene therapy protocols were also approved.

In December 1991, a conference titled 'Human Gene Therapy' was held at National Institute of Health in USA and concluded that gene therapy would soon become a potent force in medicine to deal with heart diseases, liver diseases, diabetes, and a variety of cancers. It was hoped that gene therapy will eventually play a role in disease prevention by correcting deficiencies right at birth. The cellular vehicles for gene transfer also multiplied and included endothelial cells and myoblasts, besides lymphocytes. Efforts were also made at University of California, San Diego to treat the children with AIDS using gene therapy. Stem cells from umbilical cord blood were transformed with anti-HIV gene for an enzyme cleaving HIV.

Gene Therapy Deaths and the Future of Gene Therapy : After the first case of gene therapy attempted in 1990, gene therapy was considered to be a promising area of research for human health care. But after almost a decade, in the year 1999, a solitary case of death of an 18-years old patient (Jesse Gelsinger) at the University of Pannsylvania's Institute of Human Gene Therapy, set in motion a fresh debate on the desirability of conducting clinical trials on human subjects. The patient suffered with an heritable disease of liver and a heavy dose of gene for ornithine-transcarbamylase or OTC (needed for removing excess of ammonia from blood) carried by a vector (crippled adenovirus), was introduced. The patient died on 17th September, 1999, four days after the genetically altered virus was injected into his liver.

The death was attributed to massive immuno-response caused by adenovirus and resulted in a major blow to the field of gene therapy in USA and elsewhere. All clinical trials at University of Pennsylvania's Institute of Human Gene Therapy were stopped and an investigation was ordered. Several other deaths of patients undergoing gene therapy were not reported and were sometimes attributed to causes other than gene therapy. FDA (Food and Drug Administration) of USA was therefore scrutinizing all gene therapy programmes across USA. As a consequence of the above, the use of adenovirus that was once considered to be the most efficient vector (capable of infecting both dividing and non-dividing cells and capable of expressing the gene within 24 hours, as against 3-6 weeks needed with other vectors), has been curtailed in several laboratories.

Vaccines and Diagnostics

Biotechnology has found a wide range of applications in medicine involving prevention, diagnosis and cure of diseases. Through human genetics, it has also found use in genetic counselling, antenatal diagnosis, and gene therapy. In forensic medicine, it has already been used for identification of individuals, who could be criminals (murderers or rapists). Some of these applications will be briefly discussed in this chapter under the following headings :

(i) animal and human health care,

(ii) genetic counselling and

(iii) forensic medicine. Several other aspects of biotechnology applications in medicine, including stem cells and tissue engineering are directly relevant to the use of biotechnology in medicine, and are discussed in author's another book *"Biotechnology and Genomics"*.

Animal and Human Health Care

Biotechnology may have applications in animal and human health care either indirectly or directly. Its indirect significant role involves manufacture of nutritional food through single cell proteins and in detecting food contaminants through molecular probes, etc. Its direct and more important use, on the other hand, involves efforts to overcome the menace due to a variety of diseases which can be discussed at three levels of manipulation :

(i) providing immunity against a disease through development of a vaccine,

(ii) diagnosis of a disease at early stage of its onset, and

(iii) treatment of the disease. These three aspects of dealing with animal and human diseases and 'fertility control' will be briefly discussed in this section.

Development of Vaccines for Immunity : As one of the strategies, vaccines can be prepared from animal material by isolating antigens. In conventional methods, separation of these antigens sometimes has the risk of contamination and therefore, may become health hazard. Therefore, biotechnology for development of vaccines may be used at least in four different ways, which include :

(i) separation of a pure antigen, using a specific monoclonal antibody,

(ii) synthesis of an antigen with the help of a cloned gene, and

(iii) synthesis of peptides to be used as vaccines, and

(iv) development of DNA vaccines.

Immunopurification of Antigens using Monoclonal Antibodies : Immunopurification involves separation of a specific antigen from a mixture of very similar antigens. This purified antigen can then be used for developing vaccine against a pathogen. The purification can be very effectively achieved using monoclonal antibodies, which are very specific in their reaction against an antigen. Individual interferons which are proteins having a property of inhibiting viral infection and cell proliferation, have also been purified using monoclonal antibodies. After such purification, interferons were used for clinical trials, before these were released for commercial use.

Synthesis of Antigens Through Cloned Genes : Hundreds of genes in eukaryotes have been cloned either from genomic DNA or from cDNA. These cloned genes included a number of genes for specific antigens, and in some cases have been used for the synthesis of antigens leading to the preparation of vaccines. Following two examples can be used to illustrate the use of cloned genes for vaccine preparation :

(i) Cloning of Hepatitis B virus (HBV) genome. The HBV genome has been cloned in the plasmid pBR322 and propagated in *E. coli*. From this clone, antigen could be produced in good quantity, which reacted with hepatitis B core antibody (HBA_b). This has, therefore, been used to produce hepatitis B vaccine, which was later approved for mass vaccination in several countries.

(ii) Cloning of human malarial gene. Despite efforts, no anti-malaria vaccine, could be developed so far. Recently with cloning of a gene coding for surface protein of the sporozoite of *P. falciparum*, and due to availability of whole genome sequence of this parasite, there is a hope for developing a vaccine. In human host, malarial parasite passes through several anti- genically distinct phases, namely;

   (i) sporozoite, the form in which the parasite is injected with mosquito bite; sporozoites enter the liver, multiply there and develop into.

   (ii) merozoites, which in turn invade and multiply in red blood cells; small fraction of these merozoites in red blood cells form.

   (iii) gametocytes, which may be picked up by a mosquito to start another cycle. Therefore, vaccines can be developed to control any of these phases and will be accordingly called (i) antisporozoite vaccine, which will prevent malaria in the vaccinated individual and also block its spread; (ii) antimerozoite vaccine which will protect or ameliorate the patient but will not check the spread of the disease and (iii) anti-gametocyte vaccine, which would prevent the spread without helping the patient. Of these, *antisporozoite vaccine* has been developed up to cloning of gene meant for circumsporozoite (CS) protein. This gene was obtained directly from DNA of erythrocytic form of parasite, rather than as cDNA from mRNA. The gene has been used for the synthesis of CS protein leading to the development of vaccine, which was used for clinical trials, but could not reach the commercial stage till the end of the year 2001.

Synthetic Peptides as Vaccines : Vaccines can also be prepared through short synthetic peptide chains, which have, therefore, become a subject of considerable research activity. In order to synthesize peptides to be used as vaccines, structure and function of proteins involved should be studied. Since, it is the three dimensional structure (TDS) and not the amino acid sequence, which is responsible for immunogenic response, it may be necessary to find out the protein region involved in immunogenic response.

For instance, in Foot and Mouth Disease Virus (FMDV), it is the amino acid 114-160 of virus polypeptide, which can produce antibodies neutralizing FMDV and thus provide protection. Neutralization of FMDV was also possible through the region of 201-213 amino acids of the same protein. It has thus been shown that small synthetic peptides representing these regions of proteins can show immunogenic response and can, therefore, be used for development of a vaccine.

An alternative approach to find out the immunogenic region of protein is through the study of gene coding the protein. It was also shown that cloned gene of an immunogenic protein of a pathogen (e.g., feline leukaemia virus) can be cut into fragments by DNAase I and these fragments can be cloned in lambda phage, where they may express. Phage colonies (plaques) having different cloned fragments are screened with a specific monoclonal antibody that neutralizes the pathogen. The fragments, which react with antibody must be synthesizing the immunogenic peptide fragment. This cloned DNA fragment can then be sequenced.

It was possible, in this manner to identify a 14 amino acid immunogen of the envelope protein of feline leukaemia virus (FLV). The corresponding synthetic peptide was also found to compete with the virus for antibody. When injected in guinea pigs, such synthetic peptides also elicited a partial immunogenic response. Therefore, there is a great promise for the use of such synthetic peptides to be used as vaccines. Actually a vaccine for malaria in the form of a synthetic peptide has already been prepared and is being tested for its suitability. This is the first example of a vaccine developed in the form of a synthetic peptide.

The immunogenic region of the protein of a pathogen can also be identified, by eluting it from purified MHC molecules (MHC = major histocompatibility complex). Different MHC allelic variants are available in cells for binding of different proteins and they can be purified using specific T cells. Peptides can be eluted from these purified MHC molecules and sequence of such peptides can be determined and used for manufacturing synthetic peptides to be used as vaccines.

DNA Vaccines : A DNA vaccine consists of a DNA sequence encoding the relevant immunogenic protein. The gene is isolated from the infectious organism (against which the vaccine is needed), and cloned. A promoter sequence from a eukaryote (preferably from human) is fused with the gene, to facilitate expression of this gene in human cells. This gene construct is delivered in human muscle cells and is expressed there, leading to the synthesis of immunogenic protein, which invokes both humoral and cell mediated immunity. The DNA vaccine in the form of gene construct, as above, is delivered into human body in one of the following forms : (i) direct DNA delivery by microinjection or particle gun; (ii) vector mediated gene transfer, *in vivo* or *in vitro*; when DNA is delivered *in vitro*, the genetically engineered cells are introduced into human body, so that immunogenic protein is synthesized there and immunity is invoked.

(a) *Advantages and disadvantages of DNA Vaccine :* Many of the disadvantages often encountered with other types of vaccines, discussed earlier in this chapter, are largely overcome through the use of DNA vaccines. They provide life long protection and there is no risk of vaccine causing the disease, as is the case with vaccines involving dead organisms or attenuated form of virus. The DNA vaccines are also cheap and easy to prepare, fast in action and softer to use. The only disadvantage, however, is that the gene construct that is introduced may get integrated with the genomic DNA of the vaccinated individual (animal or human) and may cause cancer. This factor alone calls for extreme precaution that needs to be exercised while using DNA vaccines for control of infectious diseases.

(b) *Present Status of DNA Vaccines.* DNA vaccines, in the recent past, were prepared using genes for several immunogenic proteins. The details of pathogens, antigens and animal systems used for DNA vaccines are listed in Table. In all these cases, generally the concerned gene was carried by vaccinia virus (vv) as a vector and the DNA vaccine thus prepared did protect the immunized animals from the concerned pathogen. Polyvalent DNA vaccines, each producing immunity against several diseases have also been developed, where several genes encoding antigens for a number of pathogens were carried by a vector (e.g. vaccinia virus).

Animal Models of DNA Vaccines.

Pathogen	*Antigen*	*Animal species*
Bovine herpes virus	Glycoprotein	Cattle, mouse
Hepatitis B virus	Capsid (core antigen)	Mouse
Hepatitis B virus	Envelope protein (surface antigen)	Mouse, rabbit, rat,
	chimpanzee	
Hepatitis C virus	Core / nucleocapsid	Mouse
Herpes simplex virus	Glycoprotein B, glycoprotein, ICP27	Mouse
Human immuno—	Envelope glycoprotein, gpl60,	Mouse
deficiency virus-1	noninfectious particles	Nonhuman primates
House dust mite	Allergen	Rat
Influenza virus	Hemagglutinin, matrix protein,	Chicken, ferrets, mouse
nucleoprotein		Nonhuman primates
Leishmania major	Major surface glycoprotein, Mouse	
Lymphocytic chorio-	Glycoprotein, nucleoprotein	Mosue
meningitis virus		
Mycobacterium tuberculosis	Hsp65	Mouse
Mycoplasma pulmonis	Hsp65	Mouse
	DNA expression library	
Papilloma virus	Major capsid protein L1	Rabbit
Plasmodium yoeli	Circumsporozoite protein	Mouse
Rabies virus	Glycoprotein	Mouse
Simian immunodeficiency	Env, gag	Monkeys
virus		
Schistosoma japonicum	Paramyosin (Sj97)	Mouse

Several of the DNA vaccines as above have been and are being tested for their efficacy in humans and livestock. However, majority of them are still in an early investigation stage and are yet to be used for human clinical trials.

Using the above four approaches, vaccines against several pathogens including the following pathogens have either been produced in recent years or are expected to be produced in the near future:

(i) Rabies virus,

(ii) Foot and Mouth Disease Virus (FMDV)

(iii) *Salmonella typhimurium,* causing typhoid,

(iv) *vibrio cholerae,* causing cholera,

(v) *Hepatitis* B *Virus,* causing hepatitis (hepatitis B-vaccine produced through recombinant DNA technique has now been approved for mass vaccination in several countries),

(vi) *Plasmodium falciparum,* causing malaria,

(vii) feline leukaemia virus (FLV) causing cancer, and

(viii) *Taenia solium* causing cysticerosis.

Diagnosis of Diseases

Diagnosis of diseases caused by protozoa and helminths. In order to supplement the conventional methods available for diagnosis of different diseases, biotechnology has provided monoclonal antibodies and DNA probes as two very effective and most sensitive tools (for techniques of monoclonal antibodies and DNA probes). Monoclonal antibodies and DNA probes are being prepared and made available for the diagnosis of a variety of diseases. Monoclonal antibodies can be used through serological tests, which will take only minutes, while the conventional methods may sometimes take weeks, since they may require culturing of bacteria or viruses as in case of herpes virus or other viruses. DNA probes, which are even more sensitive than monoclonal antibodies, will, however, take hours (not minutes as in monoclonal antibodies) instead of weeks. Readymade DNA probes for herpes virus and other human, animal and plant viruses are being prepared. DNA probe kits are also available to help preparation of DNA probes, and a market of DNA probes for several hundred million dollars (>$500 million) per year is now available. This market will expand and grow in future.

The diagnostic methods at the DNA level have the advantage over other methods, since in these methods genes of the parasite are examined and not the expressed product (used for mAbs), which changes in different stages of the life cycle of the parasite and also due to the environment. Rapid progress in recent years has been made in the development of nucleic acid based assays for the diagnosis and epidemiological surveillance of human parasites. Probes are now available for a number of human parasites from the groups protozoa and helminths (both Platyhelminths and Nematy-helminths), which are listed in following Table.

In India also, as early as 1986, a diagnostic probe for the detection of malaria was constructed at the Astra Research Centre India (ARCI), which was established as a joint venture between India and AB Astra, Sweden, (inaugurated by the then Prime Minister, Rajiv Gandhi on Jan. 7, 1987). Methods for the detection of several other diseases are also being developed at this centre.

In the above use of DNA probes, DNA has to be radioactively labelled, which in not very safe in field study. Therefore, techniques for non-radioactive labelling of probes, methods for quick preparation of target DNA and suitable protocols for hybridization are being developed.

This, in future, will allow quick diagnosis of parasites in the field.

Some of the Protozoa and Helminths where DNA Probes and Diagnostic Tests are Available.

Parasite	*Probe used*	*Hybridization technique used*
I. Protozoa		
1. *Leishmania* (kala-azar)	Kinetoplast DNA (kDNA); a repetitive DNA sequence)	Dot blots; Southern blots; *in situ* hybridization
2. *Trypanosoma* (sleeping sickness)	Kinetoplast DNA (kDNA); a repetitive DNA and total parasite DNA	Southern biot (restriction fragment length polymorphism = RFLP)
3. *Plasmodium* (malarian parasite; *P. falciparum, P. vivax and P. cynomolgi)*	Repeated DNA; synthetic oligonucleotide	Dot blots; blood samples lysed directly on nitrocellulose filters
II. Helminths		
1. *Schistosomes* (human blood flukes)	Ribosomal RNA gene	Dot blots; Southern blots
2. Nematodes; *Wuchereria* and *Brugia* (filaria) : *Onchocerca* (river blindness); *Trichinella*	Repetitive DNA sequences; synthetic oligonucleotides; ribosomal DNA	Crude DNA; blood samples; dot blots
3. Taenilid cestodes *(Taenia solium, etc.)*	Ribosomal DNA	Southern blots

Diagnosis of sexually transmitted diseases (STD) using monoclonal antibodies. *(a) Characteristics of human sexually transmitted diseases.* The occurrence of sexually transmitted diseases (STD) particularly in the developed countries and in relatively free society in our country has increased in recent years. The most common pathogens responsible for STDs are (i) *Neisseria gonorrhoeae,* (ii) *Chlamydia trachomatis* and (iii) *Herpes simplex virus* (HSV). These pathogens also have a role in a wide spectrum of diseases and therefore their simple diagnosis tests will be a great help.

(I) *Neisseria Gonorrhoeae* causes the following diseases : (i) In men : urethritis, epididymitis, (ii) In women : urethritis cervicitis, endometritis and salpingitis and (iii) In both men and women : protiditis and pharyngeal infection. In a number of infected persons, other diseases also appear which include arthritis, dermatitis, endocarditis, meningitis, chorio-amnionitis, premature delivery and blindness.

The diagnosis of gonorrhoea is convention-ally done through microscopic examination and through culture of the bacterium on a selective medium taking few days, so that biotechnological diagnosis market for diagnosis of this disease is enormous.

(a) *Chlamydia trachomatis* also causes infection in urethra, cervix, rectum and conjuctivea, and leads to diseases, some of them common with gonorrhoea. It also causes pneumonia in young infants. The clinical similarities of *N. gonorrhoeae* with *C. trachomatis* are complicated by their co-transmission in many, so that the diagnosis of *C. trachomatis* is very difficult.

In case of *Chlamydia,* the infectious form (called elementary body) enters the cell and within 48-72 hours forms a large inclusion body containing several hundred elementary bodies, which on release cause infection in neighbouring cells. When cells were fixed in ethanol (18-72 hours after infection) and stained with fluorescein-conjugated antibody, characteristic inclusion bodies were detected as early as 18 hours after infection, by immunofluorescence (IF). Detection was also possible in cultures derived directly from patients, or even on smears on microscope slides prepared from specimens derived from infected tissue of the patient. It has been shown that the direct test takes less than 30 minutes, which is a major advantage over the culture method.

(II) *Herpes Simplex Virus* (HSV) infections are most common and once caught, it remains life long and is transmitted. It is classified as HSV1 and HSV2. The diagnosis is mainly done through cell culture; type 1 and type 2 are distinguished through immunological analysis (this is called HSV typing).

(b) *Classical methods for diagnosis.* The infectious diseases such as STD are diagnosed usually by any one of the following four classical methods (these methods do not involve biotechnology).

(i) *Microscopic examination* of tissue specimens and exudates for identification of virus infected cells, bacteria, fungi or other parasites.

(ii) *Culture method* with selective growth media permitting multiplication of some micro-organisms, which can then be tested for susceptibility to potential therapeutic agents.

(iii) *Immunological methods*, where antigens associated with specific pathogens are identified in tissue or body fluids.

(iv) *Measurement of antibody* produced in the patient as a result of infection with an organism.

Laboratories often use combination of two or more of the above methods to be sure about the diagnosis, and still the tests are sometimes not full-proof, and are time and labour consuming. The use of monoclonal antibody, on the other hand, provides full-proof and very specific tests for diagnosis.

(c) *Preparation and selection of antibodies for diagnosis of STDs :* The cell hybrids are first prepared as done in hybridoma technology. Since there is a random loss of chromosomes in cell hybrids, screening of cell lines is done through replica plating technique. Hybrid cells are placed in a 96-well microtest plate. Small sample of culture fluid from each well is placed on replica plates, each of which is impregnated with a specific antigen. The immune reaction is detected through radio-immunoassay in which ^{125}I-labelled protein A, which binds to Fc protein of human IgG in the immune complex, is added to each well and then examined by autoradiography. The positive reactions can be traced back to original wells and the cell line then multiplied for production of specific monoclonal antibodies.

Using the above technique, monoclonal antibodies could be prepared that would distinguish between *N. gonorrhoeae, C. trachomatis* and HSV. Antibodies could also be selected which reacted with specific strains of the pathogen. A mixture of three antibodies (4-G5, 2-H1 and 3-C8) have been shown to identify more than 99% of the isolates of *N. gonorrhoeae,* showing no reaction with other species of *Neisseria.* The allowed diagnosis of gonorrhoea with reasonable certainty.

The suitability of monoclonal antibodies in the identification of Herpes Simplex Virus (HSV) has also been demonstrated. A panel of four different monoclonal antibodies allowed distinction between the types HSV1 and HSV2. Therefore, monoclonal antibodies can be used effectively in diagnosis and typing of HSV directly on primary clinical specimens, derived from oral, genital, mucocutaneous or ocular sites.

Antenatal Diagnosis of Congenital Diseases : When a pregnant woman is known to have a chance of bearing a child with a genetic defect, it may be desirable to diagnose the condition in the fetus. This can be done by taking some cells from the fetus by drawing a few millilitres of amniotic fluid with the help of hypodermic needle. The amniotic fluid has free cells of fetal origin, which can be cultured and tested in various ways e.g. karyotype, enzyme production and restriction fragment length analysis of its DNA. At least 35 diseases which can be identified by this technique are known. If disease is detected through such an antenatal diagnosis, abortion of fetus can then be recommended. However, if abortion is not acceptable to parents, there is no point in carrying out antenatal diagnosis.

It is possible to identify a disease now within two month's of pregnancy, unlike an 18 weeks period required earlier. The number of disease specific DNA probes is also increasing at a fast rate, so that antenatal diagnosis by DNA analysis or 'enzyme linked immunosorbant assay' (ELISA) should be possible for all single gene defects. In the past, the incidence of the disease thallassaemia, in Cypriot community in Britain had fallen from 30 to 20 per year, due to the use of antenatal diagnosis.

Treatment of Diseases

Treatment of disease involving biotechnological approaches includes development of drugs (discussed in of author's another text book "Biotechnology and Genomics", manufacture of drugs in a cost-effective manner, gene therapy and use of cultured cells and transplanation including tissue engineering discussed earlier). Some of the biotechnological approaches will be discussed in this section.

Pharmaceutical drugs through biotechnology. Drugs for treatment of diseases can also be manufactured using biotechnology. Many of the drugs for treatment are such, which could earlier be obtained only by sacrificing animal life. Two such drugs are insulin for treatment of diabetes and interferons for treatment against some tumour viruses. Such drugs can be manufactured now in bacterial cells in large quantities, if the corresponding genes from human or animals are cloned through plasmid vectors in bacteria, thus making their production relatively very cheap.

The gene for insulin was cloned in bacteria and could be used for synthesis of insulin. Dr. Saran Narang, a scientist of Indian origin, working in Ottawa, Canada was involved in cloning of insulin gene. Synthetic insulin manufactured in this manner is now being sold commercially in North America. About a dozen different interferons are also in different stages of testing and some of them are already being sold commercially. This drug 'interferon' was earlier used to be available at the rate of 16 million U.S. dollars per 50 mg costing a patient, 150 dollars per day. It is estimated that with the availability of synthetic interferons this may reduce to a cost of one dollar per day. A number of other proteins like urokinase, factor

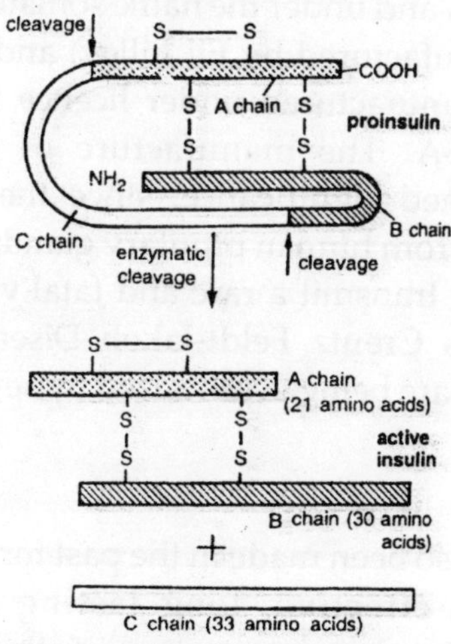

Structure of insulin and its derivation from a much longer proinsulin molecule.

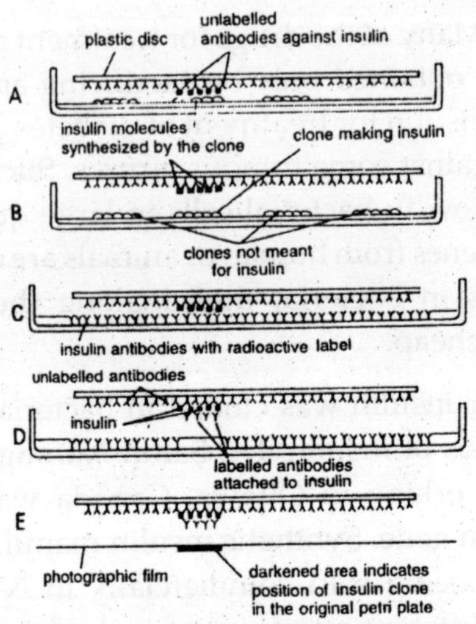

Isolation of DNA clone encoding insulin.

VII:C, human growth hormone (HGH) and many other drugs will soon be available through this technique. In 1985, human growth hormone for treating hypopituitary dwarfism was synthesized using recombinant DNA technique and was approved for commercial marketing under the name prototropin in USA and under the name somatonorm in Britain. Both insulin (manufactured by Eli Lilley) and human growth hormone were manufactured under licence from Genentech Inc. based in USA. The manufacture of *prototropin* and *somatonorm* assumed significance, since the use of growth hormone derived from human pituitary glands was halted due to fear that it may transmit a rare and fatal viral neurological disease known as Creutz Feldt-Jakob Disease (CJD). More growth hormones are being synthesized by several companies.

Fertility Control

Efforts have also been made in the past for developing safe, efficacious, cost-effective, long lasting and reversible contraceptives. In this connection, following examples demonstrate the success of Indian scientists in this area.

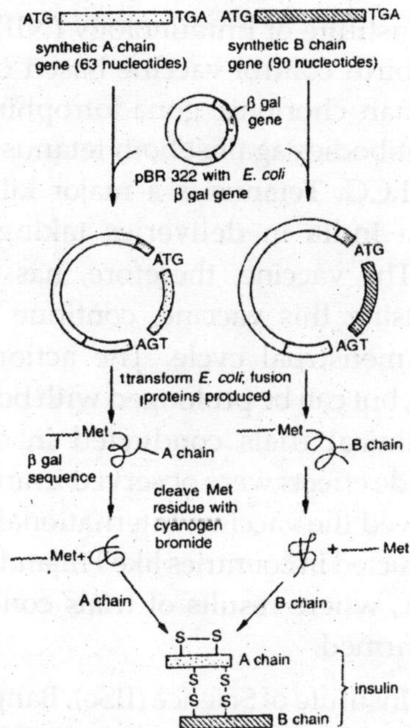

Strategy used for synthesis of insulin from cloned DNA segment.

Once a Weak Oral Pill : For 20 years, scientists at the Central Drug Research Institute (CDRI), Lucknow, worked for the development of an oral, once a week, non-hormonal contraceptive pill. This effort resulted in developing a pill called centchroman sold in the commercial name 'Saheli', which was sent to the market in June, 1991. This pill has proved to be free from any side effect and is now in great demand all over the world, being the only non-hormonal contraceptive pill.

Vaccines for Fertility Control : Immunological approaches for development of vaccines have also been tried. One project in India was initiated as a part of family planning programme by Indian Council of Medical Research (ICMR) under the Ministry of Health and Family Welfare, but was transferred to the Department of Biotechnology (DBT) in 1986. Three vaccines were being developed simultaneously :

(i) National Institute of Immunology (NII) developed the first ever birth control vaccine based on the hormone hCG (human chorionic gonadotrophin). The vaccine elicited antibodies against both tetanus and pregnancy hormone hCG. Tetanus is a major killer of pregnant women in India in deliveries taking place outside hospital. The vaccine, therefore, has double benefit. Women using this vaccine, continue to ovulate and maintain menstrual cycle. The action of vaccine is reversible, but can be prolonged with booster injections. During clinical trials conducted in early 1990s, no harmful side effects were observed in any of the subjects who received the vaccine. International trials have also been conducted in countries like Finland, Sweden, Chile, Brazil, etc., where results of trials conducted in India were confirmed.

(ii) At Indian Institute of Science (IISc), Bangalore, scientists found that administration of anti-FSH antibodies to male monkeys renders them sterile without any side effects. Studies on safety, efficacy and standardization were also conducted to undertake phase-I trials.

(iii) Another group of scientists at IISc worked on targeting vitamin carrier proteins for contraception. Lowering the concentration of vitamin carrier proteins through the use of antibodies results in rejection of embryo. After the results are established in nonhuman primates, the work would be considered for trials in humans following normal procedures.

Genetic Counselling

Chance of Having a Child with Congenital Defect : Genetic counselling for couples who believe that there may be a risk of producing a child with a congenital disease, has now become routine aspect of medical practice, particularly in the developed countries. These parents may either voluntarily abstain from having any child or may undergo selective abortions on suspicion or after ascertaining it through antenatal diagnosis as discussed earlier.

A genetic counsellor should first be able to identify the disease and therefore should be first a clinician and then a geneticist. The hereditary or developmental defects are tested through amniocentesis. At about fifteenth week of pregnancy, a sample of amniotic fluid containing fetal cells is removed and cultured *in vitro*. The cultured cells are then used for karyotype analysis and biochemical enzyme tests. The simplest cases asking for genetic counselling will be those having a family history of disease and the parents may like to know the chances of having a child free of that disease. In the simplest case, a couple may have one abnormal child (carrying disease) and would like to know the chances of having a normal child on the next pregnancy.

In such a case, if the defect is controlled by single recessive gene and both parents are normal, the chance is three in four of having a normal child, although even a normal child will have a two third chance of being a carrier. The parents may like to give birth to such a child who may be a carrier, because the chance of his or her spouse also being a carrier will be remote. However, in such cases, even the possibility of having the abnormal grand child can be worked out, if the frequency of heterozygotes in the population available for marriage of the child is known. For instance, in case of fibrocystic disease (cystic fibrosis) of the pancreas, the frequency of heterozygotes in general population (in UK) is 1/22.

The chance of the normal child to be a carrier being 1/22, the chance of the grand child to be abnormal would be 2/3 x 1/22 x 1/4 = 1/132 (since 1/4 is the chance that a child born to both heterozygote parents will be defective). A risk of one in 132 may or may not be worth taking depending upon temperament of the parents and the circumstances. More detailed calculation are possible in these simple cases and also in cases of polygenic inheritance and variable penetrance. The readers are advised to consult a book on human genetics for further details.

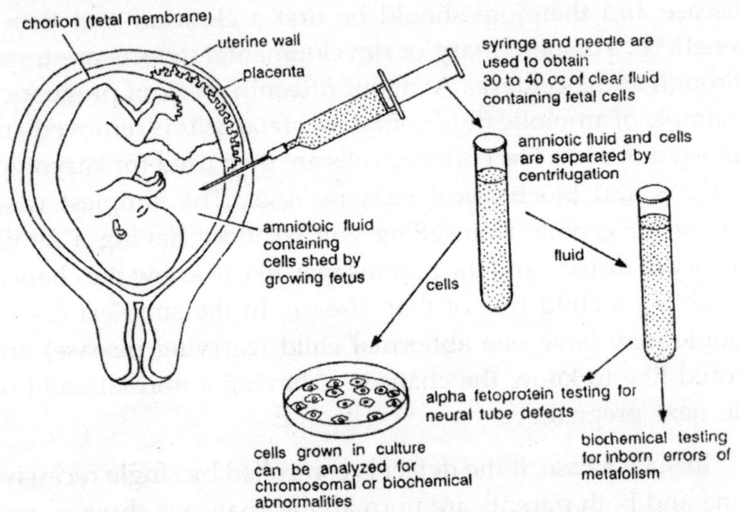

Different steps involved in amniocentesis for a test of hereditary/developmental defect.

In some cases, detection of a heterozygote may also be useful and possible. Following are the situations where it is possible, (i) When heterozygote, though phenotypically normal, produces a particular enzyme activity intermediate to those found in two homozygotes, the heterozygote can be identified. In such cases if biochemical laboratory is available, deficiency like HGPRT (Lesch-Nyhan Syndrome) in heterozygous condition can be detected from a blood sample or some skin cells. (ii) When the mutant produces an altered form of gene product, the heterozygote may produce two different forms of proteins that can be separated by electrophoresis to enable the identification of a heterozygote. (iii) If the defect is associated with a chromosome structure, then with the availability of a cytogenetics laboratory, such an abnormality in heterozygous condition can be identified.

Making a Choice of Baby's Sex

Married couples have always been interested in knowing the sex and sometimes in making a choice about the sex of their baby. The chromosome techniques have made it possible to know the sex of developing fetus by drawing amniotic fluid and preparing karyotypes from cells derived from the fetus and floating in this fluid. There are clinics available now which can advise pregnant ladies about the sex of the developing fetus, so that the ladies can decide early about the abortion of fetus, if it belongs to the unwanted sex. Techniques have also been developed, which will not require preferential abortion but will allow preferential fertilization by male (carrying Y chromosome) or female (carrying X chromosome) determining sperms.

Sperms carrying Y chromosomes, can be separated from the ejaculate of a man (through Ericson's method developed by R. Ericson of U.S.A.) to be used for insemination of ovulating women. This technique (using quinacrine stain) has been used with 80% success in more than 50 sperm centres in the world including one in Bombay at Khar Road. Ericson had actually established a company named Gametrics Ltd. in California, U.S.A. which specialized in separating sperms with Y chromosome and hundreds of male children were produced with its help.

Techniques have also been developed to separate sperms carrying X chromosome for artificial insemination leading to the birth of female children. This technique involves 'sephadex gel column' in which sperms with Y, being lighter are trapped in gel and those with X being heavier reach the bottom of the column, and can be used for insemination.

The techniques permitting choice of sex of the baby have been condemned by many sociologists, who fear that this may disturb the sex ratio leading to a variety of problems. But some doctors argue that this will help couples in planning their-families, since there are also couples who may like to have a female child. This may also allow selection against sex linked abnormalities in the children.

Forensic Medicine (Identification of Murderers, Rapists, etc.) : The technique of DNA fingerprinting developed by Alec Jeffreys and his colleagues at Leicester University in U.K. in the year 1985-86 and that of autoantibody fingerprinting described by A.K. Francoeur in the year 1988 have completely revolutionized forensic biology. In this field, identity of a person with the help of blood stains, semen (sperms) stains, hair roots, tears, saliva and perspiration will be possible with almost absolute certainty. These two techniques will be briefly discussed in this section.

DNA Fingerprinting using Minisatellite DNA : In this technique, DNA will be isolated from blood stains, semen stains or hair roots and will be subjected to Southern blotting and DNA hybridization with the help of specific DNA probes. The probes correspond to hypervariable minisatellites in DNA, each made up of tandem repeats of short sequences (11-20 bp long). A large number of these minisatellites are scattered throughout the human genome which were first detected in an intron of human myoglobin gene. This will reveal polymorphism in DNA, which has a very stable inheritance.

It is speculated that the above technique will allow identification of criminals in murder cases, rapists in rape cases and of mother and/or father in case of doubtful parentage. This technique will allow identification even when the stains on victim's clothes, etc. are several years old and with much more certainty than has hitherto been possible through techniques of blood groups, etc., since the number of blood groups available becomes a limitation. The technique of DNA fingerprinting reveals such a great polymorphism that the possibility of two persons having the same pattern of DNA fingerprints is very remote.

In India, Dr. Lalji Singh (Director, CCMB = Centre for Cell and Molecular Biology, Hyderabad) used Bkm probe and other molecular markers for DNA fingerprinting and helped in settling disputed identity in more than 50 cases. In veiw of the significance of this work, a separate institute for fingerprinting (CDFD = Centre for DNA Fingerprinting and Diagnostics) was established in Hyderabad, which is headed by Dr. S.E. Hasnain, and undertakes routine DNA fingerprinting for a variety of purposes.

Autoantibody Fingerprinting using Dipsticks : Autoantibodies that react with cellular components occur in high frequency in patients with systemic rheumatic diseases. However, a novel class of autoantibodies that react with cellular components was also identified in normal humans and other animal species. Unlike the disease associated autoantibodies, that are restricted in number, these human autoantibodies increase in number from birth up to the age of two years and then remain constant for decades, if not life long. The complement of these autoantibodies present in an individual is unique and for this reason they have been named individual specific (IS) autoantibodies. These IS autoantibodies when physically separated comprise an antibody fingerprint that can serve to identify people just like DNA fingerprints discussed above. For these autoantibody fingerprints, body fluids such as blood, semen, tears, saliva, and perspiration can be used. The advantages of autoantibody fingerprinting over DNA fingerprinting include the following: (i) sensitivity (less than 10 μl blood needed); (ii) rapidity (only few hours needed; no DNA extraction required, which may take time); (iii) simplicity (no equipment required); (iv) cost effectiveness; (v) portability; (vi) autoantibody fingerprints, though are similar in the newborn child and the mother, can distinguish between genetically identical individuals like identical twins which can not be distinguished through DNA fingerprinting.

The autoantibody fingerprinting protocol involves (i) preparation of antigen and (ii) antibody fingerprint assay. Following steps are involved: (i) A panel of antigens is prepared from an extract of human cells (e.g. HeLa cells) cultured *in vitro;* the antigens are first separated by molecular mass with denaturing polyacrylamide gel electrophoresis (SDS-PAGE); (ii) the antigens are then transferred electro-phoretically to a nylon membrane; (iii) unound sites on the membrane are blocked with the help of a blocking agent; (iv) the membranes are cut into strips called dipsticks, which are (v) incubated with dilutions of sera or plasma for 1 hour, (vi) washed with buffer, (vii) incubated in detector molecule (1 μ Ci/assay of ^{125}I protein, which binds to the Fc protein of human IgG in the immune complex formed) for 1 hour; (viii) rewashed and dried and (ix) finally used for autoradiography on X-ray film or scanned with a gamma scanner or an optical scanner. The results obtained can be utilized for the identification of individuals (humans, dogs, mice, cows, horses, rabbits, etc.)

Bibliography

Ahmad, K. Henikoff, S. : *Epigenetic Consequences of Nucleosome Dynamics,* MIT Press, Cambridge, 2002.

Altaian, S. : *Transfer RNA,* MIT Press, Cambridge, 1978.

Anderson, S. : *Sequence and Organization of the Human Mitochondrial Genome Nature,* Oxford University Press, London, 1981.

Andre, C. A. Levy and V. Walbot : *Structure of Plant Mitochondrial Genomes,* John Wiley, London, 1992.

Atkins, J. F. : *Recoding : Dynamic Reprogramming of Translation,* Oxford, New York, 1996.

Bennett, R. H. : *Transcription, Processing and Editing in Plant Mitochondria,* Oxford, New York, 1992.

Bloom, L. B. : *Biochemical Basis of DNA Replication Fidelity,* MacMillan, London, 1993.

Branden, C. and J. Tooze : *Introduction to Protein Structure,* Garland, New York, 1991.

Darnell, J. E., H. F. Lodish and D. Baltimore : *Molecular Cell Biology,* Scientific American Books, New York, 1995.

Das, A. : *Control of Transcription Termination,* Routledge, New York, 1993.

Davidson. J. N. : *Biochemistry of Nucleic Acids,* Cox and Wyman, London, 1965.

Demidov, V. V. : *PNA and LNA throw Light on DNA, Trends,* Oxford, New York, 2003.

DuPraw, E. J. : *Cell and Molecular Biology,* Academic Press, New York, 1969.

Ebby, S. R. : *Non Coding RNA Genes and Modern RNA,* Oxford, New York, 2001

Echols, H. : *Fidelity Mechanisms in DNA Replication,* Oxford, New York, 1991.

Ellis, R. J. : *Chloroplast Protein Synthesis,* Oxford, New York, 1975.

———— : *Molecular Chaperones,* Oxford, New York, 1991.

———— : *The Biogenesis of Chloroplasts Protein,* Oxford, New York, 1974.

Fischer, L.M. : *DNA Unwinding in Transcription and Recombination,* Oxford, New York, 1982.

Friedberg, E. C. : *Relationships Between DNA Repair and Transcription,* Oxford, New York, 1996.

Gesteland, R. F. and J. F. Atkins : *The RNA World,* Oxford, New York, 1993.

Gilbert, W. and B. Muller : *Isolation of the Lac Repressor,* Oxford, New York, 1966.

Gilham, N. W. : *Genetic Analysis of the Chloroplast and Mitochondrial Genomes,* Oxford, New York, 1974.

Gilmour, R. S. and J. Paul : *Tissue Specific Transcription of the Globin Gene in Isolated Chromatin,* Oxford, New York, 1973.

Glick, B. and G. Schatz : *Import of Proteins into Mitochondria,* Oxford, New York, 1991.

Gray, M. W. : *The Evolutionary Origin of Organelles,* Oxford, New York, 1989.

Grunstein, M. : *Histones as Regulators of Genes,* Oxford, New York, 1992.

Gupta, P. K. : *Cell and Molecular Biology,* Rastogi Publications, Meerut, 2000.

Hadjiolov, A. A. : *The Nucleolus and Ribosome Biogenesis,* Oxford, New York, 1985.

Hoagland, M. G. : *The Present Status of the Adaptor Hypothesis,* MacMillan, London, 1959.

Hoeijmakers, J. H. : *Nucleotide Excision Repair from Yeast to Mammals,* MacMillan, London, 2001.

Holley, R. W. : *Structure of a Ribonucleic Acid,* McGraw Hill, London, 2002.

Ingram, V. : *The Biosynthesis of Macromolecules,* Benjamin, New York, 2001.

Jacob, F. : *Genetic Regulatory Mechanisms in the Synthesis of Proteins,* Oxford, New York, 1961.

Jukes, T. H. : *Mitochondrial Codes and Evolution,* Oxford, New York, 2000.

Kaguni, L. S. : *Eukaryotic DNA Polymerase-primase : Structure, Mechanism and Function,* MacMillan, London, 1999.

Kaziro, Y: *Structure and Function of Signal Transducing GTP Binding Proteins,* Blackwell, London, 2001.

Kendrew, J. C. : *The Three-dimensional Structure of a Protein Molecule,* Sinauer, New York, 1991

Kornberg, A. and Baker, T.A. : *DNA Replication,* Freeman, New York, 2002.

Kramer, A. :*The Structure and Function of Proteins Involved in Mammalian Pre-mRNA Splicing,* Grandada, New York, 2001.

Kurland, C. G. : *Structure and Function of the Bacterial Ribosomes,* Oxford, New York, 2000.

Lake, J. A. : *The Ribosome,* Oxford, New York, 1981.

Lehninger, A. L. : *The Mitochondrion,* Academic Press, New York, 2001

Lewin, Benjamin : *Genes*, Oxford University Press, New York, 2000.

Maxwell, E. S. and Fournier, M.J. : *The Small Nucleolar RNAs*, MacMillan, London, 1995.

McKnight, S. L. : *Molecular Zippers in Gene Regulation*, Princeton, New York, 2002.

Merrick, W. C. : *Overview : Mechanism of Translation Initiation in Eukaryotes*, Longman, London, 2001.

Moore, P. B. : *Molecular Mimicry in Protein Synthesis*, Cambridge, London, 2002.

Noller, H. F. : *Ribosomol RNA and Translation*, Oxford, New York, 1999.

Olson, R. E. : *Enzymatic Synthesis of DNA*, John Wiley, New York, 2001.

Owen, L. :*DNA Synthesis*, Freeeman, San Francisco, 2001.

Pabo, C. O. and R.T. Sauer : *Transcription Factors : Structural Families and Principles of DNA Recognition*, Printice Hall, London, 1993.

Pardee, A. B. : *Molecular Basis of Biological Regulation: Origins of Feedback Inhibition and Allostery*, Arnold, London, 2003.

Pardue, M. L. and J. G. Gall : *Molecular Hybridization of Radioactive DNA to the DNA of Cytological Preparations*, Dekker, New York, 2001.

Paszkowski, J. : *DNA and Histone Methylation in Plants*, MacMillan, London, 2002.

Razin, A. and A. D. Riggs : *DNA Methylation and Gene Function*, Longman, New York, 2002.

Rich, A. and S. H. Kim : *The Three Dimensional Structure of Transfer RNA*, Oxford, New York, 1999.

Robertson, M. : *RNA Processing : the Post-RNA World*, Oxford, New York, 1999.

Rosenberg, M. and D. Court : *Regulatory Sequences Involved in the Promotion and Termination of RNA Transcription,* Cambridge, New York, 2003.

Sager, R. : *Cytoplasmic Genes and Organelles,* Academic Press, New York, 2003.

Schmid, M. B. : *Structure and Function of the Bacterial Chromosome,* Granada, London, 1990.

Spector, D. L. : *The Dynamics of Chromosome Organization and Gene Regulation,* Oxford, New York, 2005.

Stryer, L. : *Biochemistry,* W.H. Freeman & Co., New York, 2002.

Temin, H. M. : *RNA Directed DNA Synthesis,* MacGraw-Hill, New York, 2002.

Turner, B. M. : *Histone Acetylation and Control of Gene Expression,* MacMillan, London, 2002.

Ulrich, B. : *Mechanism and Regulation of Eukaryotic Protein Synthesis,* Oxford, New York, 2002.

Uptain, S. M. : *Basic Mechanisms of Transcript Elongation and its Regulation,* MacMillan, London, 1997.

Vernadsky, V. I. : *Organellar Heredity,* Revan Press, New York, 1978.

Voinnet, O. : *RNA Silencing : Small RNAs as Ubiquitous Regulators of Gene Expression,* MacMillan, London, 2002.

Wallace, D. C. : *Diseases of the Mitochondrial DNA,* Blackwell, New York, 2002.

Wang, T. S. F. : *Eukaryotic DNA Polymerases,* MacMillan, London, 2002.

Watson, J. D. : *Molecular Biology of the Gene,* Oxford, New York, 2002.

———— : *Molecular Structure of Nucleic Acids : a Structure for Deoxyribose Nucleic Acid,* Oxford, New York, 2001.

Wilkins, M. H. F. : *Physical Studies of the Molecular Structure of*

Deoxyribose Nucleic Acid and Nucleoprotein, MacMillan, London, 2002.

Williams, R. C. : *Reconstitution of Active Tobacco Mosaic Virus from its Inactive Protein and Nucleic Acid Components*, Oxford, New York, 1955.

Williamson, B. : *Split Gene Transcription*, Oxford, New York, 2001.

Woese, C. R. : *The Genetic Code : The Molecular Basis for Genetic Expression,* Harper & Row, New York, 2003.

Wuthrich, K. : *Protein Structure Determination in Solution by Nuclear Magnetic Resonance Spectroscopy*, Oxford, New York, 2001.

Yamamoto, T. : *A Bipartite DNA Binding Domain Composed of Direct Repeats in the TATA Box Binding Factor TFHD*, Oxford, New York, 2001.

Yamashita, S. : *Transcription Factor TFHB Sites Important for Interaction with Promoter-bound TFIID,* Springer, London, 2000.

Yams, M. : *Translational Efficiency of Transfer RNAs : Uses of an Extended Anticodon,* MacMillan, London, 2003.

Yanofsky, C. : *The Tryptophan Synthetase System*, Oxford, New York, 2001.

Zawel L., and D. Reinbcrg : *Common Themes in Assembly and Function of Eukaryotic Transcription Complexes*, MacMillan, London, 2002.

Zuckerman, B. : *Initiation of mRNA Translation in Prokaryotes*, Oxford, New York, 1990.

Zurick, D. : *Origin and Evolution of Mitochondrial DNA*, Oxford, New York, 1989.